Governing through Biometr

Governing through Biometrics

The Biopolitics of Identity

Btihaj Ajana

*Lecturer, Culture, Digital Humanities and Creative Industries,
King's College London, UK*

First published 2013 by
PALGRAVE MACMILLAN

Palgrave Macmillan in the UK is an imprint of Macmillan Publishers Limited, registered in England, company number 785998, of Houndmills, Basingstoke, Hampshire RG21 6XS.

Palgrave Macmillan in the US is a division of St Martin's Press LLC, 175 Fifth Avenue, New York, NY 10010.

Palgrave Macmillan is the global academic imprint of the above companies and has companies and representatives throughout the world.

Palgrave® and Macmillan® are registered trademarks in the United States, the United Kingdom, Europe and other countries.

ISBN 978–0–230–32161–8

This book is printed on paper suitable for recycling and made from fully managed and sustained forest sources. Logging, pulping and manufacturing processes are expected to conform to the environmental regulations of the country of origin.

A catalogue record for this book is available from the British Library.

A catalog record for this book is available from the Library of Congress.

Who are you? *Tu quis es.* That is an abyssal question.
(Schmitt, 1950)

Contents

Acknowledgements

In the course of developing this work, I have benefited greatly from the support and guidance of many people in ways that are hard to acknowledge appropriately here. Mostly, I would like to express my appreciation and gratitude to Nikolas Rose for his mentoring, intellectual support and generosity and to the research community at the BIOS Centre for their assistance and encouragement. Thanks in particular to Joelle Abi-Rached, Megan Clinch, Amy Hinterberger, Caitlin Connors, Lamprini Kaftantzi, Sara Toccetti, Des Fitzgerald, Susanna Finlay, Rachel Bell, Caitlin Cockerton, Kerstin Klein, Cathy Herbrand and Angela Filipe for their humour, emotional support and friendship. I am also thankful to my colleagues at the Department of Media, Culture and Creative Industries and the Department of Digital Humanities at King's College London. Special thanks to Christina Scharff, Ofra Koffman, Bridget Conor, Tim Jordan, Rosalind Gill and Andy Pratt for their continuous help and encouragement.

I thank Howard Caygill and Engin Isin for their extremely helpful comments, feedback and suggestions on an earlier version of this work. The dialogues and exchanges with friends and colleagues from Goldsmiths College and the stimulating discussions in Howard Caygill's Contemporary Thought seminars and in the Continental Philosophy Research Group have been a wonderful source of inspiration. Special thanks to Shela Sheikh for having kept me more or less up to date with the relevant interesting events and seminars. I wish to express my appreciation to Joanna Zylinska and Sarah Kember for their great and inspiring mentoring during my studies at Goldsmiths College. I also wish to thank my students at King's College London.

Very special and heartfelt thanks go to Fatima, Zoubida, Dean, Chiara, Kiarash, Christian and Lesley. Their support, warmth, presence and laughter have been a great sustenance. I also thank Michelle, Mansor, Hamid and Rami for their friendship and encouragement, and all the people the encounter with whom has shaped my thinking in many profound ways.

I also wish to thank the editorial staff at Palgrave Macmillan for their interest in this project and assistance with its publication.

Some of the research and writing of this book have benefited from the financial support of the London School of Economics and Political Science, La Fondation pour l'Innovation Politique, a grant from Funds for Women Graduate, a recognition award from International Federation of University Women, and a Travel and Research Grant from University of London Central Research Fund. I wish to thank these institutions for their generous support.

Sections of this work are adapted from the following publications: Ajana, B. (2010). 'Recombinant identities: Biometrics and narrative bioethics', *Journal of Bioethical Inquiry*, vol. 7, no. 2 and Ajana, B. (2012). 'Biometric citizenship', *Journal of Citizenship Studies*, vol. 16, no. 7.

Introduction

The tale of biometric identity systems is certainly greater than the sum of its parts. It is more than a story about 'technology', more than a story about 'identity', more than a story about 'systems' and even more than a story about their concatenation. Entwined with this tale are issues of immigration and citizenship, narratives of belonging and membership and discourses of security and threat that, in recent years, have pervaded the landscape of policy and governance and have become some of the top priorities on the political agenda of many countries. While the concept of identity systems itself is not new, nor is the application of biometric technology as such, the amalgamation of the two and the extent to which they are made amenable to rapid proliferation within all areas of society have triggered a host of concerns over the potential implications of introducing biometric identity systems. Yet, literature on the socio-political dimensions of biometrics in general and on identity-related developments in particular still remains limited to only a few studies relating to the burgeoning field of surveillance research. In addition, the majority of these studies tend to stay at one level of analysis, directing the focus mainly towards the familiar normative issues of privacy and data protection, or else towards the issue of financial costs involved in introducing such systems.

While privacy, data protection and financial costs are far from being unimportant issues, the aim of this book is to provide alternative ways for approaching biometric identity systems and for uncovering some of their multiple dimensions, multi-layered aspects and complex dynamics vis-à-vis the domain of governance in general. What concerns me primarily is how forms of identity, citizenship and belonging, and how modes of identification, are increasingly being redefined and reconfigured in the name of risk and security, and made amenable to various

1

modes of securitisation and control through technology. In the course of this book, I argue that some of these transformations and processes unfold in the field of *biopolitics*, that is, the management of life whereby the merging of body and technology, together with the segmentation of society, is precisely what is at stake. Two overarching and illustrative figures are being juxtaposed in this book, namely the 'neoliberal citizen' and the 'asylum seeker'; this, in an attempt to uncover the aporetic and complex character of such contemporary forms of biopolitical governing, and to demonstrate how each of these two figures implicates the other.

Invoking the notion of biopolitics inevitably leads us to invoke the notion of bioethics. For where there is a politics of life, there is also an ethics of life (assuming here the inseparability of politics and ethics in the Lévinassian fashion). As such, this study also represents an enquiry into the *bioethical* implications of biometric technology and identity systems, techniques that are largely based on the biological characteristics of human bodies. Pinning the prefix 'bio' to 'ethics' here is by no means an attempt to simply follow the current hype of bioethics whereby the notion of the 'bio' itself is, as Cooper et al. (2005) argue, at risk of becoming as 'expansive a term as Marx's concept of social reproduction – a black box where everything that had previously been discarded from economic [moral] and political philosophy is conveniently recuperated'. Instead, my attempt is that of approaching the ethical implications of biometric technology in terms of *embodiment* and *singularity*, concepts that challenge both the traditional framework of bioethics (with its prescribed cluster of codes, regulations, principles and protocols) and the libertarian approach (with its taken-for-granted values of liberty, privacy and autonomy). The suggested alternative, in this book, starts from the bottom-up proposition that 'ethics is bodies' (Thacker, 2004: 188) – that it is first and foremost a question of (embodied) 'ontology' (Nancy, 2000). So in this sense, the inclusion of the 'bio' in '*bio*ethics' is a matter of emphasis rather than hype, and biometrics, as a biopolitical field of knowledge and power, provides us with a valid site for rethinking bioethics anew while uncovering some of the workings of biopolitics itself.

In what follows, I shall introduce some of the key concepts and issues around which the chapters of this book are organised. I also consider, in this section, some aspects relating to the epistemological framework adopted in this book, highlighting some of the reasons as to why an *experimental* approach is needed to attend to the multi-layered biopolitical dimensions of biometric technology and identity systems. One valuable contribution of this work is, in fact, its innovative approach, which operates at various levels – theoretical, empirical, political, ethical and thematic – in ways that reveal the explicit and implicit links between them. It is an approach that seeks to challenge and critique contemporary theoretical positions from the vantage point of empirical material while at the same time drawing on these positions to critically reflect on that material and shape its theorisation.

Biopolitics of biometrics

Biometrics, which is literally the 'measurement of life', refers to the technology of measuring, analysing and processing the digital representations of unique biological data and behavioural traits such as fingerprints, eye retinas, irises, voice and facial patterns, body odours, hand geometry and so on. It can be used in two ways: *identification* in order to determine who the person is, through one-to-many comparison, and *verification* in order to determine whether the person is who he claims to be, through one-to-one comparison (Mordini and Petrini, 2007: 5). The emergence of biometrics as a 'popular candidate' (Lyon, 2003: 667) for identification and authentication systems is mainly due to its ability to automate the process of linking bodies to identities; distribute biological and behavioural data across computer networks and databases; be adapted to different uses and purposes; and (allegedly) provide more accurate, reliable and hard-to-tamper-with means of verifying identity. Like other (traditional) identification systems, the procedure of biometric identification consists of four stages: *enrolment* (digital representations of unique biological features are captured through a sensor device, and then processed through an algorithmic operation to produce a 'template'), *storage* (the produced template is stored on a database or/and on a chip card), *acquisition* (as with the enrolment stage, a biometric image is captured and transformed through similar algorithmic procedures into a 'live template') and *matching* (the live template is compared to the stored template to establish whether the person is known to the system, in the case of database, or whether the live biometric capture corresponds to the one on the card, in the case of chip card) (European Commission, 2005a: 35).

Beyond these basic technical definitions, biometrics can also be defined as a form of 'new media' to the extent that it digitally *mediates* between the body and identity, between technology and biology. Chapter 1 explores this alternative definition by drawing on Thacker's concept of 'biomedia' and Bolter's and Grusin's notion of 'remediation'. Through these concepts, the chapter seeks to highlight two key elements: first, the centrality of the body and embodiment to biometric techniques and how this affects the very ontological foundation of biometrics as well as that of the body itself. Here, I argue that the biometric body acts, at once, as a 'medium' and the 'mediated' through its interface with biometric technology, while the latter represents something that exceeds its status as a mere instrument or extension. The second element refers to the relation of biometrics to its predecessors. Instead of assuming the novelty of biometrics and subscribing to the current hype around it, this chapter traces its genealogy by considering some older techniques of identification that have been deploying the body as their mode and object of measurement, and how biometrics *remediates* these techniques and *refashions* their socio-political significance. Establishing a historical continuity between biometrics and its ancestors, such as anthropometry and fingerprinting, helps us understand the conditions of possibility and the historically embedded contexts framing and underlying the emergence of biometrics.

In trying to theorise and understand the biopolitics of biometrics, it is necessary to understand first the concept of biopolitics itself and its manifold variations. As such, Chapter 1 also provides a thorough discussion on biopolitics by considering the work of three canonical theorists of the politics of life, namely Foucault, Agamben and Rose. The connections, divergences and contentions in their accounts on biopolitics offer a rich theoretical framework through which one can explore biometrics as a technology of biopower whereby the body and life itself are the subject of modalities of control, regimes of truth and techniques of sorting and categorisation.

Function creep and social sorting

Some of the major concerns surrounding the developments in biometric identity systems revolve around the phenomena of 'function creep' and 'social sorting'. It is often argued that the procedural spread of biometric techniques across different spaces of governance signals towards the quantitative and qualitative increase in surveillance practices and the intensification of technology's power to profile and sort through individuals and populations. Stalder and Lyon (2003), for instance, argue that just as paper-based identity documents have been historically used to mark certain people as members and single out others for different treatment,[1] so too is the case with the current high-tech identity systems. Especially, following the events of September 11 and other attacks, certain groups were constructed as a threat to national security and became an easy target to various sorting and profiling activities within America, Europe and other parts of the world. People of Arab or Muslim background, appearance or names, in particular, were immediately affected by the backlash effect and the climate of suspicion resulting out of these attacks (ibid.: 88). So the fear is that the linking of identity systems to searchable databases that contain individuals' unique bodily and behavioural characteristics may contribute to the automation of modes of social sorting and ultimately the legitimisation of discriminatory practices in a more implicit and sophisticated way, all within the rationality of 'pre-emptive' surveillance.

For Stalder and Lyon, mechanisms of anticipatory and pre-emptive surveillance represent a shift towards a 'New Penology'. Unlike the 'Old Penology', which sought to identify 'individuals' in order to ascribe guilt, blame and modes of punishment, this New Penology instead seeks to implement techniques of identification and classification in order to manage groups according to their level of perceived dangerousness – although older models of power 'can kick in at any time' (ibid.: 90). It is part and parcel of what Levi and Wall (2004: 200–1) call 'soft security through surveillance', which relies on proactive identification of 'risky groups' and suspect populations.

Certainly, engaging with the notion of social sorting from the perspective of 'risk', 'suspicion' and 'security' is crucial to understanding some of the implications of introducing biometric identity systems. In Chapter 2, however, I argue that this approach may also miss the point that something more complex is taking place vis-à-vis the classificatory character of biometric technology than simply electronic sorting on the basis of ethnicity, religion and so on. This is because biometric identity systems can also *perform* the function of classification 'the other way around', that is, to mark certain people as holders of 'a *surplus* of rights', to use Balibar's (2002: 83) evocative expression. For instance, in various airports across the world, frequent flying executives may avoid long queues at passport control by using chip cards containing their digital information or by simply scanning their irises (see also Chapter 4). In this sense, what is needed is a more inclusive approach that recognises the *polysemic* character of biometric identity systems with regard to social sorting.

Furthermore, Chapter 2 also attempts to go beyond the seemingly technologically determinist debates on biometrics' functional creep. While these debates have a tendency to narrowly focus on the *technical* and operational character of the biometric spillover across different spaces and practices (thereby precluding other concerns), here I frame the issue instead from a wider *political* perspective taking into account the *embodied* ramifications of biometrics. By drawing on Agamben's take on biopolitics in terms of the notion of 'exception', I explore how exceptional and illiberal measures and policies are increasingly being deployed in the field of border management and immigration control. This notion of exception is largely accredited to Carl Schmitt (1985), for whom 'the sovereign is he who decides upon the exception'. It indicates the array of imperatives, conditions and situations that necessitate responses and measures that go beyond the limit of what is deemed as 'normal' politics and are driven instead by unconstrained sovereign power. Agamben takes Schmitt's claims rather seriously and sees in them the horizon of what has become of the contemporary political landscape. So much so that exception is becoming the norm, according to Agamben. Correlatively, I argue that the function creep of biometric identity systems can be addressed in light of their spillover from exceptional spaces (e.g. criminal justice, immigration and asylum policy) to the general body of humanity and in terms of their becoming a normative and all-encompassing practice. Nevertheless and as demonstrated by a set of empirical examples throughout the book, I stress that this 'totalising' spillover of biometrics does not affect everyone in the same way, nor has the notion of exception become somewhat generalisable and homogenous. Instead, this expansionary move is underlined by a polysemic and multi-layered feature that facilitates the selection of those to be surveilled and the normalisation of the rest.

Identity, informatisation and embodiment

The implementation and rapid spread of biometric systems inevitably call the status of identity itself into question. More specifically, and given its fundamental characteristics, biometrics raises a need for exploring and understanding the intimate and intricate relationship between identity, body and information. To be sure, the informatisation and digitisation of the body together with notions of de/materialisation and dis/embodiment are some of the issues that have dominated much of the (techno)cultural, sociological, scientific and philosophical debates since the late twentieth century. And with the current development and proliferation of biometric technologies, these issues are given renewed importance, rendering the body once again a central point of discussion and bringing the problematic bifurcation of information and materiality into sharper focus. Some of the seminal works in this area relate to studies of Irma van der Ploeg (2003a, 2005a) in which she seeks to rethink the entire normative approach by which the confluence between body and technology is understood and conceptualised, and reveal the extent to which the distinction between 'embodied identity or physical existence [...] and information about (embodied) persons and their physical characteristics' can be sustained (ibid.: 2003a: 58).

The ways by which the body is transformed into processable, storable and retrievable information are numerous and among the most notable ones are the techniques of genetic fingerprinting, DNA typing and the growing field of bioinformatics. In all of these techniques, what is enabled is the process of acquisition, storage and analysis of biological information via algorithmic and computational methods whereby new forms of knowledge production are generated and in which the notion of 'body as information' is salient.

This ontology of body as information construes the body itself in terms of informational flows and communication patterns,[2] exposing the porous and malleable nature of body boundaries. And when the body is viewed beyond its somatic and material contours, what ensues is a problematisation of the very distinction between materiality and immateriality and, with it, the distinction between the 'material' body and the body as 'information'. This, in turn, poses a challenge to 'issues previously considered self-evident' so much so that the 'presumed demarcation of where "the body itself" stops and begins being "information" will subtly shift [and] the moral and legal vocabularies available will no longer suffice' (ibid.: 67).

In this respect, and especially with regard to the normative concerns of privacy and bodily integrity relating to biometrics and biotechnologies in general, van der Ploeg argues that the 'legal' and 'ethical' distinction between what is perceived in the dichotomous discourse as 'the thing itself' (the body),[3] and the digital representation of that 'thing' (i.e. the personal information held on the body),[4] is likely to flatten as a result of this ontology of body as information – this, despite the continuous upholding of the difference between searches *on* the body and searches *in* the body. Examples of the problematic efforts to deal with the increasingly blurring boundaries between 'the body itself' and 'body as information' can be seen, for instance, in the controversy surrounding the much-contested field of DNA sampling and banking of genetic information. In these practices, it is not the act of touching the body or crossing its anatomical–physical boundaries that is at issue vis-à-vis the normative notion of bodily integrity. Rather, it is the taking of a DNA sample itself even though the methods of doing so can hardly be noticeable to the person involved. As such, *what* is stored about a person is 'constitutive of, and inseparable from' *who* that person is. And according to van der Ploeg, this argument demands a serious and urgent rethinking of what is at stake in the 'intensive forms of monitoring, categorizing, scrutinizing and, ultimately, controlling and manipulating of persons through their bodies and embodied identities' (ibid.: 70–1).

These concerns are also present in Alterman's (2003) analysis of biometrics' implications vis-à-vis embodiment and identity. He contends that the spread of our bodily 'representations' across networks and databases entails a fundamental 'loss of privacy, and a threat to the self-respect which privacy rights preserve' (ibid.: 143). He distinguishes between what he calls *biocentric* data (e.g. biometric data) and *indexical* data (e.g. social security number, driver's licence number and so on). While the former is centred on the 'body', the latter has no 'internal relation to an embodied person; it possesses no property that is tied to our psychological or physical conception of self' (ibid.: 144). Such a distinction allows Alterman to posit that indexical data have no intrinsic relationship to 'one's dignity or self-respect', whereas biocentric data have a direct impact on 'one's right to control the use and disposition of one's body' (ibid.: 145).

Taking cue from Kant's formula that 'one must treat people only as ends in themselves, never merely as a means', Alterman argues that the use of a person's body as a means to an end is akin to making that person surrender her body and with it her 'free will'. And in the context of biometric identification, this amounts to *objectifying* the body and 'providing it as a means to an end in which the person has no inherent interest' (ibid.). He thus concludes that biometric identification produces a sense of 'alienation' from one's body as well as from the technology that is used to identify it (ibid.: 146). In so doing, it jeopardises the privacy rights needed to preserve the 'psychological comfort zone' by which the 'public self' is able to maintain itself (ibid.: 143).

In a sense, one cannot help but feel sympathetic towards Alterman's and others' privacy concerns with regard to the proliferation of biometric technology. Yet, the presumed clear-cut and taken-for-granted distinction between the 'public' and the 'private' seems rather oversimplifying. It is often argued that one of the outcomes of the developments in (bio)technology is the blurring of boundaries between what is perceived as public and what is regarded as private.[5] Of course, this argument itself is also open to contestation. However, its merit lies precisely in the way it poses certain challenges to the set of dichotomies upon which Western modern thought is premised, including the binary of private/public. Furthermore, the reliance on the Kantian moral principles of autonomy and free will to discuss the 'inherent moral value' of biocentric data runs the risk of reducing the argument to the mere notion that biometric technology 'objectifies the body by isolating the physical element from the person' (ibid.: 145). While this argument is valid to some extent, it does not, however, account for the myriad dynamics entailed within the interface between technology and the experience of embodiment – for there is more to this interface than the simple objectification of the body through technology. Also, in seemingly attributing the 'entire' agency to technology, such arguments run the risk of falling yet again into the trap of 'technological determinism'. In this determinist view (biometric), technology is 'constructed as having a stable and knowable ontological status, and as being endowed with specific properties and inherent features, that subsequently exert influence, generate impact [...] on an external world' (van der Ploeg, 2003b: 89). Again, while this view is not to be immediately dismissed insofar as it exposes the characteristics and stakes of the 'technological itself', there is a need, however, to consider other 'possible scenarios, contingent uses, and courses of actions' (ibid.). And the same can be said about the other side of the pole, that is, 'social determinism'. In this respect, the challenging task is to attend to the 'distributed agency' of biometric technology and its varying uses and functions (see also discussion in Chapter 1).

The above are but some of the contentions and debates that inspired the enquiry undertaken in Chapter 3. In this chapter, I interrogate the ways in which biometrics is *about* the uniqueness of identity and the *kind* of identity biometrics is concerned with – this, by way of furthering and deepening the debate on identity and its biometric mediation and delimiting some of the distinctive bioethical stakes of biometrics beyond the familiar terms of privacy, data protection, information integrity and the like. I draw primarily on Cavarero's Arendt-inspired distinction between the 'what' and the 'who' elements of a person, and on Ricoeur's distinction between the 'idem' and 'ipse' versions of identity. By engaging with these philosophical distinctions and concepts, and with particular reference to the case of asylum policy, I seek to examine some of the bioethical issues pertaining to the practice of biometric identification. These issues relate mainly to the paradigmatic shift from the biographical story (which for so long has been the means by which an asylum application is assessed) to bio-digital samples (which are now the basis for managing and controlling the identities of asylum applicants). Such purging of identity from the narrative dimension lies at the core of biometric technology's overzealous aspiration to accuracy, precision and objectivity, and raises some of the most pressing bioethical questions vis-à-vis the realm of identification. To respond to these issues, I propose a bioethical framework based on the 'narrativity' thesis, arguing for the inclusion of the narrative dimension of identity while assessing asylum claims instead of the reductionist approach of biometric identification. This framework, I argue, has the potential to enable a more compassionate engagement with each refugee story than what the systemacity of biometrics allows.

Securitised citizenship

The introduction of biometric systems, for the purpose of securitising identity, controlling borders and managing social services, is often regarded as being symptomatic of the transformations taking place within the domain of citizenship. In the age of economic globalisation, migration and the putative erosion of nation state, citizenship is seen as becoming a 'hollowed out' concept whose carcass is increasingly shaped around the techniques of 'identity management' and the proliferating use of biometric technology in which the body is treated as 'password'. As Muller (2004: 280) argues, the move towards the governance of identity through biometric technology transforms citizenship practices into 'a quest for verifying/authenticating "identity" for the purpose of access to rights, bodies, spaces, and so forth' as well as the purpose of (allegedly) identifying 'risky individuals'. Consequently, biometric technology is reconfiguring the relation of the individual to the polity in terms of 'security', all the while maintaining and facilitating the traditional exclusionist politics of citizenship. The shift towards identity management is, therefore, regarded as being simultaneously *depoliticising* and *politicising*.

One way of understanding the depoliticising aspect of identity management is by looking at the growing trends of *securitisation* whereby the notion of citizenship is drawn away from 'the spaces of conventional politics, towards sites of private authority and governmentalities' (ibid.: 281). Indeed, in one sense, the privatisation and governmentalisation of citizenship retract security itself from the realm of 'state politics' (although not entirely) and places it within the realm of individual 'subjectivities' and, on a more general level, within the realm of the 'societal'. For security is also, as Burke (2002: 22) suggests, a 'technology of subjectivity [that is] both a totalizing and individualizing blackmail and promise'. To this, we can add the Copenhagen School's notion of 'societal security' in which security is conceptualised in terms of social identity. As cogently explained by Williams (2003: 518),

> The concept of societal security is designed to highlight the role that 'identity' plays in security relations. Here, it is not the territorial inviolability ('military' security) or governmental legitimacy and autonomy ('political' security) that is threatened. Rather, it is the identity of a society, its sense of 'we-ness', that is at stake.

Some criticism has been directed at this notion of societal security. For one thing, its defining of society in terms of identity is underlined by the presupposition that society has a 'single' identity to be securitised, which thereby overlooks the hybridity and multiplicity of social identities (McSweeny in Williams, 2003: 519). Second, there is the risk of inadvertently legitimising – or at least not opposing – forms of intolerance and hostility on the basis of defining security as a 'defence' of *the* individual and social identity (ibid.). Added to this is the fact that 'the concept of societal security embodies the old state/society dichotomy, a formulation which fails to do justice to the mutability of political space and the inventiveness of power' (Walters, 2004: 240). But suffice to say, in this instance, that with this *socialisation* of risk and threat, it is not surprising that the metamorphosis of citizenship into identity management is being clothed in security rhetoric and adorned with biometric technology. These depoliticising trends can be witnessed, for instance, in the current hype surrounding 'identity theft' (Stasiulis, 2004: 296) reinforced by 'private' corporate institutions such as international banks and insurance companies. As a result, a new hybrid consumer-citizen is emerging and the boundaries between private and public domains are increasingly blurred. But alongside the depoliticisation of security and citizenship, there still remains the 'political' need to 'identify' the continuum of risk and threat. And it is in this process of identification that the politicising aspect of identity management comes into play.

Invoking the work of Carl Schmitt, for whom the concept of 'the political'[6] is primarily the antagonistic discrimination between friend and enemy, Muller contends that the securitisation of citizenship and identity entails 'the resilience of political agency and the aggressive politics of inclusion/exclusion' (Muller, 2004: 281). The move towards identity management, in this sense, does not entirely purge citizenship from its all too familiar binary distinctions between 'us' and 'them', 'friend' and 'foe', 'legitimate' and 'illegitimate', and so on. Instead, the introduction of biometric technology into identity governance reorganises these distinctions around the measuring, evaluation and identification of what is (potential) threat and what is not, who is legitimate and who is not, through mechanisms of digital profiling. For as Derrida (in ibid.: 284) suggests, the existence of the political itself relies upon 'the *practical identification* of the enemy'.[7] So although the 'constructivist approach' to security (seen, for instance, in the notion of societal security) may seem to be representing 'the social' as the main hallmark of securitisation, 'the political' subsists throughout the myriad processes of identity management by which antagonistic distinctions are upheld and sovereign 'decisions' are made.[8] It is for a similar reason that Williams (2003: 520) maintains that '[s]ecuritization can never be reduced to the conditions of its social accomplishment: it is an explicitly political choice and act'. It 'marks a decision, a "breaking free of rules" and the suspension of normal politics' (ibid.) in favour of 'exceptional politics' *à la* Schmitt, which makes it all the more politicising.

However, the politicising function of identity management through biometric identification does not consider 'identity' itself in the traditional way identity constituents (race, origin, gender, etc.) are linked to conventional practices of citizenship.[9] Instead, the politics of identity management is more about issues of access to resources, services, spaces and privileges. It is concerned with discriminating between 'qualified and disqualified bodies' (Muller, 2004: 290). This rationality of dis-identifying identification, so to say, is manifest in a number of sites, with immigration and asylum policy being a major one. The governance of these fields is not so much interested in the *identity* of the migrant or asylum seeker per se (their cultural background, biographical stories, etc.), but merely in scanning their bodies for signs of (il)legitimacy (those signs are themselves a construction of that same governmentality) through biometric technology in order to grant or restrict their access and construct risk profiles.

In this regard, the Schmittian distinction between friend and enemy takes another twist: 'we need not *know* the friend, but merely authorize access to particular resources, rights and entitlements to the authenticated friends, while blocking access to the unverifiable' (Muller, 2004: 286). Nevertheless, this should not be taken to mean that identity management is devoid of instances of discrimination that are linked to race, ethnicity and so on. But the subtle difference between identification (through identity) and authentication is precisely what allows biometric technology to conceal the 'cultural and ethnic attributes of citizenship' (ibid.: 280) behind its preoccupation with authenticity, and thereby hide its exclusionary and discriminatory character.

Both of these depoliticising and politicising aspects are taken up in Chapter 4, which asks the question as to what kind of citizenship is the 'biometric citizenship'. By drawing on the governmentality thesis and in examining a range of examples, from identity theft and fraud to security techniques at airports, this chapter explores the multidimensional nature of biometric citizenship. It argues that this citizenship is at once a 'neoliberal citizenship', a 'biological citizenship' and a 'neurotic citizenship' that straddles various modes of governing including *governing through freedom, governing through mistrust* and *governing through affects*.

The neoliberal aspect of biometric citizenship is demonstrated through the rearrangement of the experience of border crossing in terms of the neoliberal ethos of choice, freedom, active entrepreneurialism and transnational expedited mobility. At the same time, these are enacted alongside the exclusionary and violent measures directed at those who are constructed as risky categories, illustrating the constitutive relationship between the 'biometric citizen' and its 'other'. As regards its biological aspect, biometric citizenship is embedded within rationalities and practices that deploy the body not only as a means of identification but also as a way of sorting through different forms of life according to their degree of utility and legitimacy in relation to market economy. This aspect also carries a racial and national dimension exemplified in both identity card schemes and the very technical infrastructure of biometric technology. The neurotic dimension of biometric citizenship has to do with the way in which fear and distrust, among other affects, are constantly mobilised as a means of legitimising security measures and constructing hyper-vigilant subjects. What these three features have in common is the increasing technocratisation of citizenship and, with it, the weathering away of the very ideal of political community.

Ontology of community

Community is indeed the key theme of Chapter 5. This chapter provides a critical reflection on what transpired from the previous chapters by adopting a somewhat deconstructive approach to the notion of citizenship and that of community and placing the debate on biometric technology in a wider context that goes beyond both the governmentality thesis and the framework of exception. At first glance, this chapter may seem quite remote from the rest of the chapters in that it transposes the debate to the philosophical realm of ontology. Yet, this remoteness is rather *performative*. It stages a gesture of *retreat* in order to create a space for thinking *otherwise* about the issues raised throughout this book. So rather than staying at the level of governmentality, its practices, discourses and rationalities, this chapter engages with the *meaning* of the political itself outside the limit of political practices and seeks to rethink some deeply rooted categories underpinning Western politics and its approach towards belonging, otherness and difference.

The chapter is largely informed by the work of Jean-Luc Nancy concerning the question of community and the retreat of the political, and finds in it some helpful directions for alternative forms of collectivity that are not based on fear, control and security-driven policies (expressions of which are found in biometric identity systems). Through Nancy's co-existential analytic of 'being-with', 'being-in-common' and 'being singular plural', the chapter argues for a concept of the political that is more open to the uncertainties of the future, more accepting of difference and diversity, and for an onto-normative framework that stresses the importance of relationality over technicality, of communication over fear and of openness over prudentialism.

Governing

> Governing is a genuinely heterogeneous dimension of thought and action – something captured to some extent in the multitude of words available to describe and enact it: education, control, influence, regulation, administration, management.
>
> (Rose, 1999: 4)

This book takes the *heterogeneity* of governing quite seriously. Since heterogeneity, by virtue of its nature, is bound to yield contradictions and juxtapositions that do not always sit comfortably with one another, the overall approach of this book, towards the issue of governing, is one that seeks to engage rather than avoid contradictions, if not even, incoherences. It is, therefore, less about declaring allegiance to a specific stance, less about 'theoretical camp sitting' (Mythen and Walklate, 2006: 394) and more about *experimenting* with various approaches and 'fabricating some conceptual tools that can be set to work [or *unwork*] in relation to the particular questions that trouble contemporary thought and politics' (Rose, 1999: 5). As stated above, the main rationale of this book is to provide new ways and different options for exploring the biopolitical and bioethical dimensions and implications of biometric identity systems, and how these systems can help us understand the transformations occurring at the level of governance. At the same time, this book is also driven by a desire to challenge and supersede the simplistic and unfruitful, yet obdurate, divisions between the philosophical and the empirical, and to turn this work itself into an ethical space of *hospitality* for accommodating different ways of thinking and doing (or rather, 'thinkingdoing'): a space of conceptual contamination, cross-fertilisation and new alliances in which conflicting concepts exist in a creative tension – 'not in a liberal fashion, where each position is equally valid, but rather in a state of mutual transformation' (Zylinska, 2005: 7).

Governing through biometrics is, doubtless, a heterogeneous process that combines a myriad of practices and thoughts. Understanding these requires us to examine the rationales, discourses, meanings, dynamics and narratives involved in biometric processes. At one level, this can be captured through the governmentality approach, which since its inception by Michel Foucault, has been serving as a useful analytical tool for exploring the complex relationship between government and thought, between technologies of power and their underpinning political rationalities. Here, the term 'government' is not restricted to the political field but is broadly defined as 'the conduct of conduct' which ranges from 'governing the self' to 'governing others' (Fimyar, 2008: 5; Lemke, 2002: 2). The approach of governmentality, as such, is generally focused on questioning what and who can be governed, who can govern, through what mechanisms and for what objectives, and on diagnosing the 'styles of thought' by which certain issues are made thinkable and amenable to governmental strategies and interventions (Petersen, 2003: 191; Rose, 2007).

Crucial to the rationalities of governing is the idea of 'intelligibility'. Rose (1999: 28) argues that '[i]t is possible to govern only within a certain regime of intelligibility'. As such, language is regarded as an integral component of the knowledge grid that is produced through the different styles of thought insofar as it acts not only as a tool for describing governmental activities but more so as a condition of their possibility. For language 'does not merely "represent" pre-formed interests or aspirations: it has an active role as a kind of intellectual apparatus for rendering reality thinkable' (Lentzos, 2006: 460). In addition to its discursive character, governing is also a matter of demarcating the spaces and sites to be governed and turning them into an object of thought and the target of political programmes. In this way, space can be modelled as both an abstract entity through the distribution of cognitive topographies and a concrete one through the implementation of conceptual schemas into real spaces and sites.

In this book, I focus on specific sites that are increasingly being governed by biometric technology, including borders and citizenship (and by extension, immigration and asylum). Through the governmentality perspective, the book explores the myriad discourses through which these sites are represented as domains of governing. Parenthetically, and as Petersen (2003: 191) argues, the governmentality approach does not represent a singular 'theory of governance' per se nor a general thesis, but rather refers to an analytics of power that is best described as 'a "zone of research" rather than as a fully formed product or thesis'. It is with this spirit in mind that I deploy the analytics of governmentality, that is, in a somewhat loose and experimental way, which is neither restricted by the styles of previous studies nor circumscribed by specific research techniques.

The discourses by which biometric identity systems have been proposed and debated, legitimised or refuted, are primarily discourses about identity and otherness, community and belonging, legitimacy and illegality, entitlement and exclusion, security and threat. These are, therefore, the themes that pervade most of my analysis and reflections throughout the book. However, the examination of the issue of biometric identity systems as an *example* of contemporary forms of governance cannot be reduced to its discursive dimension (i.e. merely in terms of the linguistic construction of risk, threat, otherness, identity and so on). It also has to be considered in terms of other non-discursive practices straddling the whole paraphernalia of material techniques, bureaucracies and fields of knowledge and expertise (government, industry, media, campaigning groups, etc.) (Aradau, 2001; Bigo, 2005a; Huysmans, 2006). That is to say, the 'technological' aspect of governing.

According to Rose (1999: 52), '[t]echnologies of government are those technologies imbued with aspirations for the shaping of conduct in the hope of producing certain desired effects and averting certain undesired events'. They subsume an array of techniques, instruments, institutions, apparatuses, personnel and all other elements that contribute to making rationalities operable and enabling the execution of governmental strategies. As such, analysing the technological dimension, in the context of this book, requires a consideration of how the rationalities underpinning the governance of borders, immigration, asylum and citizenship are made functional through the deployment of biometric identity systems, and how different institutional entities and expert bodies contribute to rendering governmental rationalities not only debatable but also instrumentalisable and ultimately translatable into the domain of reality.

Taking these two dimensions (discourses and technologies) of governmentality together leads us to regard the discursive and the non-discursive aspects of biometric identity systems as a sort of continuum in which the discursive constitutes one of the conditions of possibility for the materialisation of the non-discursive (in the sense that language does not only make reality thinkable but also contributes to the justification, mobilisation and actualisation of certain material practices), just as the non-discursive is what concretises the discursive (by imbuing the linguistic element with a material effect). In this sense, the discursive and the non-discursive cannot be separated from one another but need to be regarded as parts of an *assemblage* – 'a whole matrix of corporeal and incorporeal relations [...], a complex deployments of bodies and machines as well as an order of events, discourses, concepts and formulae' (Bogard, 2006: 103). In this sense, biometric identity systems need to be regarded not just as technological artefacts or acts of discursive demarcations, but as 'functional hybrids' (Hier, 2003: 400) whose unity is based on the amalgamation of technical procedures which assign properties to bodies, and truth statements which embed these procedures into a system of knowledge and power – and back again.

As will be discussed in Chapter 1, there exist some differences and diversions in opinion and approach as to how the concept of biopolitics is theorised and deployed for analysing issues of governing. This is often cast in terms of a sharp contrast between the Foucauldians (governmentality) and the Agambenians (sovereignty). While the Foucauldian approach rejects certain philosophies of history in favour of genealogical methods that begin with the 'practices' in order to examine the contingent conditions of the co-emergence of specific projects and technologies of government (see Foucault (2008 [1979]), the Agambenian approach is more concerned with history of concepts and adopting a paradigmatic approach in which the 'origin' of a concept ends up determining, in a rather general and all-embracing way, its ensuing meaning, function, purpose and scope. In this book, I do not wish to declare allegiance to either. Instead, I attempt to force-marry the two approaches with one another in order to shed some light on the complexities and paradoxes that permeate current rationalities and technologies of governing. What interests me epistemologically, in doing so, is to primarily elucidate how the paradigm of exception (and with it the notion of sovereignty), of which Agamben speaks, is not completely removed from the field of governmentality or set in opposition to it, but constitutes in itself a specific form of governmentality – something I demonstrate in Chapter 2 via Didier Bigo's elaborations. For, as Fimyar (2008: 10) argues, following on from Stenson, 'it is misleading to separate technologies of governmentality from discipline and sovereignty, because they are "not equivalent entities". Instead it is more productive to perceive "governmentality as a broad framework of governance, within which discipline and the sovereign control over territory operate simultaneously"' (Stenson, 1999: 54).

Overall, the value of using the analytics of governmentality for approaching the issues surrounding biometric identity systems lies in its ability to provide a diagnosis of the hybrid arguments, strategies and modalities of thought and action that underpin such mechanisms, and to open up a space for their critique. It also lies in its ability to reveal how specific forms of subjectivity and modes of (de)subjectification come into being through the different governmental practices and within their various management sites.

Despite its benefits, however, the governmentality approach has its own limitations. These are summarised cogently by Petersen, who points out that the selective focus on 'abstracted' rationalities and technologies of rule limits the potential of governmentality studies to contribute to change and radical politics (Petersen, 2003: 197–8). This also risks turning governmentality literature into '"ritualized and repetitive accounts" of "governing" in increasingly diverse contexts' (O'Malley in ibid.: 198). For such reasons, Petersen follows many other commentators in advocating an expansion in the methodological repertoire of the governmentality approach 'beyond the history of the present and a focus on the mentalities or rationalities of rule' and the cultivation of 'a more fruitful dialogue with other kinds of social science' (Petersen, 2003: 198–9).

For my part, I take issue mainly with the 'superficial' character of the governmentality approach. While some governmentality theorists explicitly advocate superficiality, or to put it more precisely, 'an empiricism of the surface' (against interpretation and hermeneutics of depth) (see, for instance, Rose, 1999), I see in remaining at the surface a rather unsatisfying gesture and a danger of restricting one's understanding and critical thrust. This is why in Chapter 3, and in addressing the question of identity, I shift towards a more philosophical approach that hinges on the 'narrativity' thesis as formulated by Cavarero and Ricoeur. This, as will be seen, is not only a matter of epistemological choice, but also a means by which I seek to raise some pertinent ethical questions vis-à-vis the implications of biometric identification and emphasise the need to attend to the phenomenological, embodied and narrative aspects of identity that often escape the radar of governmentality studies, given the emphasis of the latter on remaining at the surface and engaging with abstract subjectivities instead.

My other concern is to do with the question of 'being'. Studies of government often focus on interrogating 'the problems and problematizations through which "being" has been shaped in a thinkable and manageable form [...], the techniques and devices invented, the modes of authority and subjectification engendered, and the telos of these ambitions and strategies' (Rose, 1999: 22). While this is all worthwhile, in focusing only on the panoply of technologies and rationalities by which 'being' has been shaped, the question of 'being itself' is left in abeyance. In fact, this focus may as well unwittingly contribute to instrumentalising being even further and uncritically accepting its increasing and excessive technologisation, to the extent that studies of governmentality tend to rely on the very notions and elements they intend to critique. Chapter 5 turns, therefore, to 'ontology' instead to address precisely this question of being and its implications with regard to the notion of the 'political' and that of 'community'. This is done through an exploration of Jean-Luc Nancy's approach to these questions, details of which (including some epistemological considerations) will be discussed in that same chapter.

In a way, then, if I am using the governmentality perspective in this book, I am only doing so as *un enfant terrible* who is not afraid of wandering and engaging in methodological promiscuity. And it is in this sense that I have deployed the conceptual toolkits offered by the various approaches, somewhat inventively, experimentally and at times even riskily. This has helped retrieving that which has been left behind by the governmentality approach (namely 'narrativity' and 'ontology'), while, at the same time, engaging with that which is usually omitted from the philosophical approach (e.g. 'practices' of governing).

It is worth mentioning, at last, another dimension that has contributed a great deal to my epistemological approach and to the shaping of the ways in which I select, formulate and engage with the questions raised throughout this book. This relates to the subjective dimension. As someone who grew up in a non-Western country, I have borne *witness* to how the ontology of 'papers' can truly be a matter of life and death; how what is taken for granted in the 'West' and is denied to the rest can become an internalised fetishistic object of desire with a value

and a force of its own; and how unsettling it can be the sense of alienation and non-belonging afflicted on certain groups through the power
of state documentation and its attendant ramifications. This is partly
the backdrop behind my ethico-political drive to reveal some of the violent aspects that inhere to technologies of security, identification and
borders in general.

Like a shaman, I have journeyed throughout this research, visiting
along the way some of the deepest cellars of my mind, absorbing at the
very cellular level many of the affects emanating from the figures that
have been summoned in this research: I got in touch with the uncanny
refugee inside me, with the neoliberal subject I had to become and with
the singular being I am. Each of these embodied subjectivities (though
I prefer not to regard the latter as a form of 'subjectivity' for reasons
that shall be divulged in Chapter 5) has provided me with a unique
looking glass through which I was able to comprehend, elucidate and
connect with the distinctive concerns and particular issues that figures,
such as those of the immigrant, the asylum seeker, and the neoliberal citizen, engender in their own different but nonetheless interrelated ways.
There is, therefore, a considerable, though implicit, degree of empiricism involved in this book. It is an empiricism without too much of
the empirical – an empiricism that hinges on the personal that is also
a collective, on the singular that is also plural (this will become clearer
when discussing Nancy's notion of the singular plural and being-with).
This subjective dimension, in my view, together with the situatedness,
relatedness and embodiment it entails, is precisely what preserves the
singularity of research, creates a space for intellectual hospitality and
takes the ethico-political injunction beyond the systematic choreog-
raphy of established theories – acts that are, I believe, crucial to our
rethinking of what a research on biopolitics and bioethics is or what it
ought to be.

1
Biometrics: The Remediation of Measure

This book is about the biopolitics of biometrics. Its first chapter will, therefore, address these two components as a way of laying the ground for the remaining chapters and explicating the key theoretical framework that underlies this project. The first section of the chapter is primarily concerned with defining 'biometrics'. Instead of limiting the term to its technical definitions, this section proceeds by placing biometrics within a historical context and highlighting some of its genealogical referents that have also been historically involved in measuring the body for identification purposes. The second section of the chapter turns the discussion towards the concept of 'biopolitics'. It provides a critical overview on the origin and development of this concept with particular reference to the works of Michel Foucault, Giorgio Agamben and Nikolas Rose whose different accounts inform much of the internal workings and theoretical framework of this book.

Remediating measure

Apart from biometrics' technical definitions, outlined in the 'Introduction', there are other ways by which one can define and, at the same time, historicise and thereby problematise what biometrics *is*. For instance, biometrics can also be described as a form of 'new media', or more precisely, as a form of 'biomedia' that transforms the body into machine-readable codes while also encouraging the 'biological-as-biological'[1] (Thacker, 2004: 7). Biomedia, as Thacker (2004: 13) proposes, are 'not simply about "the body" and "technology" in an abstract sense'. They rather indicate a *situated* 'interdisciplinary cross-pollination (biological computing, computational biology)' (ibid.) whereby the biological and the technological are seen to be 'mediating' each other: the biological 'informs' the technological and the technological 'corporealises' the biological. This situatedness that is characteristic of biomedia is precisely what makes the latter more than a concept and a technology, but the 'conditions' in which both 'the concept (recontextualising the biological domain) and the technology for doing so (e.g. bioinformatics tools) are tightly interwoven into a situation, an instance, a "corporealization"' (ibid.). In formulating biomedia as such, Thacker is attempting to take the argument beyond two familiar tropes: first, beyond the limitation of technology to the notion of the 'instrument', the 'tool' in which the essence of technology is considered as that which comes from the 'outside' and remains distinct from the body; second, beyond the McLuhanite take on media technology as being the 'extension' of man, a functional supplement to the human body. In opposition to these tropes, and also beyond the utopian simulacra of bodily displacement/replacement, biomedia, according to Thacker, 'do not so much configure technology along the lines of instrumentality [...] although an instrumentalization of the biological body is implicit in the practices of biotechnology' (ibid.: 14), nor do they articulate a unilinear and dichotomous relationship between the technology and the body wherein one can function as the substitute or the extension of the other. Instead, they gesture towards an irreducible interconnectedness between the two that supersedes the 'juxtaposition of components (human/machine, natural/artificial)' (ibid.: 7). Here, Thacker is seeking to problematise the dividing slash that stands between these categories, and which has, for so long, served as a means of *organising* the Western humanist thought.

Of course by now, and with all the developments that have taken place within the realm of techno-science and other related fields with regard to the problematisation of the relationship between the biological body and technology, Thacker's attempt may hardly seem unconventional. Yet, the 'thirdness' of instrumentality and extension to which Thacker is alluding through the formulation of biomedia is more than a mere reiteration of the same familiar debates and problematisations. It rather underscores a crucial and unique feature that hinges mainly on the question as to how biomedia conceive of the body – technology relationship. Thacker's answer suggests that the singularity of biomedia lies in the fact that, to some extent, biomedia do *not* conceive this relationship – they are not concerned with fixating (at least not once and for all) what may be entailed by the dash separating 'body' and 'technology'. Instead, they are more interested in establishing conditionalities and facilitating operativities that are intrinsically ambivalent, contingent and situated. This way, the biological does not disappear into the technological, nor does the technological remain purely technological:

> In biomedia, the biological body never stops being biological [...];
> it is precisely for that reason that the biological body is inextricably
> 'technological'. This does not, of course, mean that there are no tech-
> nological objects involved, and no techniques or specific practices.
> Quite the contrary. But it is how those techniques and technologies
> are articulated in these biotechnological practices that makes this a
> unique situation.
>
> (ibid.)

So, by regarding biometrics as a biomedium, we are led to consider this biotechnology less as a tool and more as a *process*, less as an instrument and more as an *act* through which various techno-bodily mediations come into being. Moreover, this process of mediation through biometrics also puts into question the status of the 'body itself'. For rather than emphasising an 'external' mediation between body and technology (as is the case with other (bio)technologies whereby techniques and processes are applied *to* the body from the 'outside' – for example, piercing, tattooing and cosmetic surgery), biometrics renders the body itself as both the 'medium' (the means by which 'measurement' is performed) and the 'mediated' (the 'object' of measurement). In doing so, biometrics creates the 'zone of the body-as-media' (Thacker, 2004: 10) where the biological and the technological are merged together.

Body-as-media

In *Remediation: Understanding New Media*, Bolter and Grusin (1999) illustrate how the body functions as a medium and is itself subject to mediation. They locate this dual process in what they term 'remediation'. Although their illustrations focus mainly upon specific technologies, such as bodybuilding and cosmetic surgery, which seek to reshape the exteriority of the body while rendering the latter as the medium of aesthetic expressions, their overall suggestion remains a case in point vis-à-vis the technology of biometrics as well. According to the authors, 'a medium is that which remediates. It is that which appropriates the techniques, forms, and social significance of other media and attempts to rival or refashion them in the name of the real' (ibid.131: 65).[2] Correlatively, '[i]n its character as a medium, the body both remediates and is remediated. The contemporary, technologically constructed body recalls and rivals earlier cultural versions of the body as a medium' (ibid.: 238).

In this sense, for Bolter and Grusin, neither the body nor technology can be considered in isolation insofar as they are both embedded in their institutional milieu (socio-cultural, economic, historical, etc.) and maintain a relation of *continuity* towards their earlier versions. The 'real' in the name of which the process of refashioning takes place is nothing other than the myriad collections of social, cultural, psychological and political arrangements and experiences that remediation (of body and technology) seeks to capture and respond to. At the same time, however, neither the 'materiality' of the body nor the 'technicality' of technology can disappear into their social constructions. By adopting such a view, the authors take us all the way back to some earlier debates on media technologies, precisely those relating to the epistemological clash between the *technological determinism* of Marshall McLuhan (in which agency is entirely attributed to technology) and the *social constructionism* of Raymond Williams (in which social needs, practices and purposes are seen to be the primary drivers behind technological change and developments). Nevertheless, Bolter and Grusin's restaging of such polarised debates is not meant to valorise or debunk either of them – as this often leads to the reductive discourse of 'all or nothing' as Kember (2006) argues (i.e. either technology is assigned too much agency or it is deprived from it altogether). It is rather an attempt to find a 'third way'[3] for rethinking and conceptualising the 'networked' relationship between the technological, the social and the body, quite apart from binary determinist attitudes:

> In an effort to avoid both technological determinism and determined technology, we propose to treat social forces and technical forms as two aspects of the same phenomenon: to explore digital technologies themselves as hybrids of technical, material, social, and economic factors.
>
> (Bolter and Grusin, 1999: 77)

It is precisely this *hybridity*, which the authors regard as an inherent feature in technology, that may add a crucial caveat to our definitions of biometrics and rescue the debate from both forms of determinism. For it is not enough to define biometrics only in terms of its technical aspect, nor is it enough to merely attend to its social and political constructions. Instead, it is necessary to consider how biometrics-as-biomedium remediates prior technologies of identification as well as prior social and cultural contexts. So let us now dwell, for a moment, on the concept of biometrics as remediation.

In locating biometrics within the notion of remediation, a fundamental question arises to the forefront: what is 'new' about biometrics? With the current hype surrounding biometrics, one often encounters a tendency of overstating the novelty of this technology and ignoring its complex history. There is, of course, something 'seductive' about the rhetoric of newness – or even 'the no-longer-newness of the new' (Kember, 2006), which is why many theorists associate the 'new' with the ideological narrative of Western progress and its attendant utopian/dystopian discourses (see, for instance, Baudrillard, 1983, 1990; Bolter and Grusin, 1999; Kember, 2006; Lister et al., 2003). This seductiveness is, in part, what conceals the *genealogy* of new technologies and obscures the historical *continuity* connecting them to older technologies. For these reasons, Bolter and Grusin (1999: 14–5) deflates the rhetoric of newness by maintaining that 'new media are doing exactly what their predecessors have done: presenting themselves as refashioned and improved versions of other media [...] What is new about new media comes from the particular ways in which they refashion older media and the ways in which older media refashion themselves to answer the challenges of new media'. That said, one can argue that to address the newness of any given technology in terms of remediation is to articulate its relation to its ancestors both within a non-teleological historical perspective and beyond paranoid or utopian futuristic discourses. To do so with regard to biometric technology and identity systems in general, we need to first understand some of the mechanisms underlying the logic of identification, why and how identification became so intrinsic to the working of modern states and why it continued to infuse the ongoing technological attempts to fix identity to the singular body (Gates, 2005: 37–8).

Identification

According to Caplan and Torpey (2001: 1–2), 'universal systems of identification are unthinkable without mass literacy and an official culture of written records'. They, therefore, relate the origins of individual identification techniques in Europe to writing itself and, hence, to the early medieval epoch that witnessed a radical transition from 'oral' to 'written' procedures.[4] With the rise of this writing and recording culture, the documentation and registration of individual identities began to be established as an official mechanism for facilitating various transactions, such as taxation, bookkeeping and property ownership, and for monitoring the movements of travellers. By the sixteenth century, 'new documents of origin and identity came to be demanded as a matter of course from ever expanding groups of people' (Groebner, 2001: 16).

The origins of individual identification can also be linked to modern concepts of individuality, subjectivity and personhood that are central tenets of the Western humanist narrative (Caplan and Torpey, 2001: 2; Groebner, 2001: 16). To this end, the principle of individual identification cannot be separated from that of 'individual identity'. Equally, '[i]dentification as an individual is scarcely thinkable without categories of collective identity' (Caplan and Torpey, 2001: 3). Here, therefore, lies in the intimate and intertwined relationship between the logic of individual identification and the establishment of collective membership and citizenship rules. Membership, as a principle, is initially constructed and performed through modes of inclusion and exclusion whereby identity is conceived of in terms of dichotomies of self and other, of inside and outside, of belonging and alien and so on. These dichotomies are the means by which sovereignty attempts to resolve the tensions embedded within the dialectics of the universal (the 'human') and the 'particular' (the 'individual' belonging to a particular state), and provides the organising principle for individuated citizenship (Coward, 1999: 6). This is because 'the citizen is the individual [and] individuality is both an expression of a claim to ontological universality and of ontological particularity' (ibid.: 5). As such, creating reliable systems of identification in which individuals are distinguished from each other and assigned fixed identities became a necessity for all modern states and an essential aspect of their 'state-ness' (Torpey, 2000: 3). Yet, and as Gates (2005: 38) argues, '[o]ne enduring problem has been that of articulating identity to the body in a consistent way'. First, the 'hybridity', 'instability' and 'changeability' of both identity and the body make it difficult to accurately connect the two together. Second, the problem of human 'fallibility' and lack of objectivity makes the process of identification by human agents rather inefficient and unreliable. And, third, the monolithic amount of archival and administrative procedures needed to operate an effective identification apparatus exceeds the capacities of even the most organised systems (ibid.: 38–9). Therefore, governments and institutions have resorted to technology in order to control individual identities in the most accurate way.

Anthropometry

Some of the earlier and most notable examples of technologies of identification can be found in the developments that took place during the nineteenth century. The rationale behind these technologies was to create a criminal history by which the state could distinguish between first-time offenders and 'recidivists', and respond to the challenges posed by the increasing migration of individuals and the rapid urbanisation of cities (Cole, 2003: 2–3). Developed in the 1880s by the French law enforcement officer, Alphonse Bertillon, anthropometry is held to be 'the first rigorous system for archiving and retrieving identity' (Sekula, in Gates, 2005: 41). As Kaluszynski (2001: 123) explains, 'anthropometry was not simply a new weapon in the armory of repression, but a revolutionary technique: it placed identity and identification at the heart of government policy, introducing a spirit and set of principles that still exist today'.

Anthropometry involved the measurement and documentation of individual bodies as well as the organisation of an identity storage system. It proceeded in two stages: *description* and *classification* (ibid.: 125). The first stage was based on the measurement of specific dimensions of the body (including height, head length, head breadth, left middle finger length, left little finger length, left foot length, left forearm length, right ear length, cheek width, etc.), which was also supplemented with a detailed and meticulous description of physical features, especially those of the face and head (Finn, 2005: 24).[5] The second stage involved the recording of these measurements onto a standardised printed card and dividing each of them into small, medium and large categories. The completed cards were then indexed and filed according to which group they fell into so that when faced with a suspect, the police could record the obtained measurements onto a new card and compare them with existing ones for possible matching (Cole, 2003: 4). However, Bertillon's system was only a means of negative identification, that is, 'a method of elimination that could prove *non*-identity' (Kaluszynski, 2001: 126). So, on its own, it was unable to achieve the forensic certainty it strived for. Therefore, Bertillon used 'photography' as a complementary procedure for creating the *portrait parlé*, a comprehensive identification card that allowed the personalisation of anthropometric data and the identification of criminal subjects (Finn, 2005: 24; Kaluszynski, 2001: 126). By merging these techniques together, Bertillon's system was able to bring more 'criminal' bodies under surveillance, establishing itself as a popular apparatus for prisons and police departments not only in France but worldwide (Kaluszynski, 2001).

Fingerprinting

Yet the success of anthropometry was merely a short-lived one as it was soon to suffer a deadly blow with the arrival of a new and a more accurate identification technique: fingerprinting. As noted by Finn (2005: 25), the origins of fingerprinting go back to the work of Henry Faulds (in Japan, during the late 1870s) and William Herschel (in India, beginning in 1850s) who both tried to trace heredity through the examination of fingerprints. Although their attempts were unsuccessful, they both realised the potential use of fingerprints as unique identifiers in criminal investigations. Like Bertillon's system, however, the use of fingerprints for criminal identification was equally prone to the same problems of organisation and classification resulting out of the large volume of data and inscriptions. So, the challenging task that was facing Faulds, Herschel and their contemporaries was to find mechanisms for transforming the fingerprint into a viable and complete system of identification. In 1891, the scientist Juan Vuccetich managed to develop what Simon Cole (ibid.: 26) refers to as 'the first classification system which rendered fingerprints a practical means of indexing a large criminal identification file'. And, a year later, Francis Galton proposed a tripartite classificatory method that divided fingerprint patterns into three types: loops, whorls and arches. Galton's method was soon superseded by the work of Edward Henry and his colleagues who extended Galton's classification system by assigning loops and whorls to subcategories (based on the distinction in the ridge characteristics of the print) (ibid.). Henry's scheme enabled the classification of individuals according to pattern types and subtypes, and the sorting of 'even very large collections of identification cards into relatively small groups' (Cole, 2003: 8).

In this sense, fingerprinting seemed to have overcome the hurdles that impeded Bertillon's system. For in contrast to the complexity and laboriousness entailed within anthropometric techniques, fingerprints provided a simpler, cheaper, faster and a more practical and accurate way of identifying individuals: '[m]ore than just a visual, numeric or textual representation, they presented a literal, physical trace of the body' (Finn, 2005: 27). So, by the 1930s, fingerprinting fully replaced Bertillonage and gained universal acceptance as being the most prominent, valid and adequate method of personal identification, whether in terms of law enforcement or in terms of other practices of civil identification (ibid.).

At this point, note ought to be taken with regard to the social and cultural contexts of the emergence and development of these two technologies. Importantly, it should be borne in mind that anthropometry and fingerprinting were enlisted not only to identify individuals but also to 'diagnose' disease and criminal propensity, define markers of heredity and correlate physical patterns with race, ethnicity and so on.

In fact, they were both tied to 'biologically determinist efforts to find bodily markers of character traits like intelligence and criminality' (Cole, 2003: 4). Such efforts were complicit with discriminatory discourses and racist practices. In France, for instance, anthropometry has served a pivotal role in 'the organization of the French Republic's mixed system of "security and repression"' (Kaluszynski, 2001: 129). It was mobilised to target and control specific groups such as 'gypsies' and 'nomads' who were ostracised on the basis of their 'excessive' mobility:

> The [police] squads were required to photograph and identify 'at any time when this is legally possible vagrant nomads and romanies traveling individually or in groups, and to submit to the supervisory authorities photographs and descriptive identifications, taken according to the anthropometric method.' Here 'for the first time files were established based on categories attached to "racial characteristics".' Itinerance itself became a 'pre-offense,' as witness a 1905 investigation that categorized gypsies according to whether they were sedentary or nomadic.
>
> (ibid.: 131–2)

Anthropometric nomad passbook

The surveillance practices were further reinforced through the introduction of the 1912 nomad law, which was a culmination of mounting anxieties expressed by public opinion, politicians and the press vis-à-vis 'undocumented' travellers. The law gave mayors juridical power over local legislation on temporary sojourn, and with it the full authority to decide whether or not to grant gypsies the right to camp within the territory of their commune (ibid.: 136). From its outset, the '1912 law implicitly took "racial indicators" into account. The nomad was regarded as an element in the population who was distinguished by his allegedly criminal otherness, and was not regarded as worthy of citizenship' (ibid.: 137). And, to tighten control over the movement of these marginal groups even further, an anthropometric pass was also introduced. The pass, which was called *Le carnet anthropométrique des nomades* [anthropometric nomad passbook], was more than an identity card but a collective record stating the physical characteristics of each member of the group as well as details of marriages and births, and other health-related information (such as vaccinations). It had to be presented to the police on arrival and departure, and on request for criminal investigations. Falsifying or failing to complete the passbook was punishable by very heavy fines including the seizure of possessions.

'Your papers!'

I showed some scraps of paper that were torn and dirty as a result of folding and unfolding.

'What about your card?'

'What card?'

I learned of the existence of the humiliating 'anthropometric card'. It is issued to all tramps and stamped in every police station. I was thrown into jail.

(Jean Genet, 1965: 75)

The imposition of these burdensome measures inflicted a great amount of constraint on the mobility of gypsies and nomads, and increased their stigmatisation. Some were even forced to give up their itinerant way of life as a result (Kaluszynski, 2001: 137). This case demonstrates clearly the 'function creep' of anthropometry, for although it was initially intended to fight recidivism, anthropometry also revealed itself as 'a technique of republican government addressed to society at large, containing the issue of access to citizenship at its heart' (ibid.: 138).

Worth mentioning as well that some of the experimental sites for the development of identification technologies in the nineteenth century were partly colonies (this being particularly true of fingerprinting). The 'body of the non-western other has played a foundational ontological role in the development of western medical and scientific epistemologies', according to Pugliese (2010: 42). In India, for instance, the bodies of local people were used to try out and master identification techniques before they were exported to the metropoles (Leonardo, 2003: 103) – just as the current biometric systems were initially trialled in exceptional spaces (such as detention centres) and on people with the 'least rights' (such as asylum seekers and prisoners) before spilling over to the rest of the population (Fuller, 2003a).[6] As Rabinow (1996: 113) points out, '[t]he first practical usage of fingerprinting took place in Bengal [...] The proverbial "prevalence of unveracity" of the Oriental races provided the motivation for these gentlemen [colonial officers] to perfect a reliable identification system, one whose basis lay in a marker beyond or below the cunning will of native or criminal'.

ID cards

The chief principle of a well-regulated police is this: *That each citizen shall be at all times and places, when it may be necessary, recognized*

as this or that particular person. No one must remain unknown to
the police. This can be attained with certainty only in the following manner: Each one must always carry a pass with him, signed by
his immediate government official, in which his person is accurately
described. There must be no exception to this rule.

(Fichte, 1889 [1796]: 378–9)

Like biometrics, national identity cards schemes also have genealogies
that are marked by complex histories and whose specificities differ
according to their geographical and political contexts. These histories
can play a key role in the reactions towards current ID card schemes:
'An ID card may be carried with pride, indifference, reluctance or even
fear, depending on the political conditions and the history of using such
documents in the country in question' (Lyon, 2009: 3). For instance,
in France, the carrying of an ID card is accepted as a matter of course
since 1940s (ibid.), while in Brazil there has been a strong support
for ID schemes out of fear of exclusion and the risk of disappearance
(Wood and Firmino, 2010). For other countries that are less familiar
with national ID cards or those that only had them during wartimes,
the reactions are rather different. The UK example is a case in point.

In addition to the recently abolished UK biometric identity scheme,[7]
'ID cards have been introduced twice in Britain, during the First and Second World Wars, but dismantled both times soon after for interesting
reasons' (Agar, 2001: 101). It is fair to say that, historically and at least
in comparison to other European contexts, there has always been something quintessentially distinctive about the debates on identity cards in
Britain; something to do with a sense of 'Britishness' itself, which began
to be imbued with life and meaning during the war:

> [A]s J.M. Winter has recently argued, the Great War threw some
> aspects of 'Britishness' into sharp relief. In particular he points to
> a celebration of the 'character' of the British soldier (first the pri-
> vate Tommy and only later the officer – middle-class, patriotic,
> unemotional, unintellectual, and masculine), but also a process of
> differentiation: what was 'English' was defined in opposition to what
> was taken to be German: decency versus bullying, fair play versus
> atrocities, amateurism versus militarism.

(ibid.: 103)

Such processes of cultural differentiation and national identity construction carried over interestingly and problematically into the entire realm

of administrative policies creating an uncomfortable tension between the state's will to document identities and the desire to preserve the distinctiveness of the so-called British character. For while the creation of bureaucratic mechanisms including regimes of identification was seen by the government as a necessity in times of war, there was also a need to reassure the public that such mechanisms would not lead to Prussification that was so characteristic of the German bureaucracy and, by opposition, 'un-British'. Yet, this tension did not stop the government from introducing an identity card system akin to that which was in operation in Germany – although the differences between the two countries continued to be upheld and emphasised throughout parliamentary discussions:

> Such a system could only be successful when enforced, as in Germany, by rigorous and ubiquitous police system upon a nation accustomed to be regulated in all minor matters of life. Any system of registration which is intended to operate successfully in this country must be based on different principles.
>
> (*Memorandum on the NR scheme*, July 1, 1915, in ibid.: 105)

In 1915, a National Register Act was passed through Parliament. It made it compulsory for everyone between the ages of 15 and 65[8] to register their personal details and inform the authorities of any changes in home address and so forth. Once a form was completed at a local registration office, the person was given a certificate to sign and keep. This certificate was the first official identity card in Britain. So, for the first time, 'British people would be known not in aggregate but by unit' (Agar, 2001: 104). Nevertheless, with the growing number of records (due to the movement of people and the reapplication for new cards), the maintenance of the National Register became rather cumbersome and useless. And by 1919, the registration system was no longer in operation.

Civil servants, however, together with the military remained keen on preserving some form of national registration and individual documentation for various purposes: 'it would appear that the present register, with all its defects, has proved itself to be invaluable for existing recruiting systems [...] if put on a permanent basis it would [...] undoubtedly justify itself as an addition to our national institutions' (*Memorandum on the National Register*, May 31, 1919, in ibid.: 106–7). And again: 'the electoral registration system, the census organisation, and the births, marriages and deaths registration system should be amalgamated so as to provide a uniform system of registration for all purposes' (the

Committee of Imperial Defence, in ibid.: 107). Despite this zealous interest in maintaining a system of registration, a peacetime National Register was ruled out on the basis of being unfeasible.

It was the outbreak of the Second World War that gave officials yet another opportunity for re-implementing a national registration system in Britain. More specifically, it was the need for food rationing during the war that provided the main justification for a second universal and mandatory identity card and secured the public's acceptance of it. So, by September 1939, a national identity card system was already in place. Although the initial purpose of the system was to ensure the fair distribution of food, it became quickly integrated into other bureaucratic and day-to-day functions (withdrawing money, opening an account, applying for passport, collecting parcels, etc.). The system was also used to track down war deserters and those evading compulsory national service, and to counter bigamy and fraud. At the same time, the police were gaining more power, through the Defence Regulations, to detain anyone not in possession of an identity card (ibid.: 108–9).

With the end of the war in 1945, resistance against identity cards and national registration started to increase. Keeping this system in peacetime, Britain was deemed unacceptable by both the public and the media:

> Except as a wartime measure the system is intolerable. It is un-British [...] It turns every village policeman into a Gestapo [...] It can put the law-abiding citizen in the same row of filing cabinets as the common thief with a record.
>
> (*Daily Express*, March 12, 1945, in Agar, 2001: 110)

> Identity cards have put another weapon into the hands of many minor officials to badger the innocent public. They have outlived their usefulness. Let's be done with them. Give us back a little bit of our traditional freedom.
>
> (Hendon and Finchley, *Times and Guardian*, June 15, 1951, in ibid.)

The government's strategic response was to connect the rationale of peacetime identity cards and national registration to the notion of welfare, and more precisely, to the provision and management of health care. But despite this attempt, the survival of the identity system lasted only until February 1952 when it finally collapsed (Agar, 2001: 109–10).

In conclusion, by placing the technology of biometrics and identity systems within a historical context, it is possible to demystify the hyperbolic novelty that is often attached to them and to reveal their close relation to older techniques and technologies. What is, in fact, new about biometric identity systems is the way they 'refashion' their predecessors, not only in the technical sense (by performing, extending, reworking and enhancing similar sets of functions and measurements), but also in terms of the socio-political context of their introduction. As can be deduced from the above discussion and examples, biometric identity systems strikingly remediate similar (but not the *same*) anxieties, motives, rationalities, functionalities, discourses, responses and so on that have shaped much of the *raison d'être* and development of anthropometry, fingerprinting, national registration and paper-based identity cards (crime control, migration, security, management of public services, etc.). This, while *maintaining* the body itself as the site of remediation whereby the biological and the technological are brought even closer. That is not to say, however, that *nothing* has changed. Instead, such conclusion merely stresses the need of keeping a genealogical perspective in mind while approaching the issue of biometric identity systems and when questioning their novelty. For there are, undoubtedly and as will be shown later on, irreducible differences in terms of the contexts in which current identity systems are being *framed* and mobilised.

Biopolitics

Having considered some of the genealogies of biometrics in the previous section, I shall now proceed to address the concept of 'biopolitics', which, as the title suggests, constitutes the main theoretical framework within which this book is placed.

Since its inception, the notion of biopolitics continued to incite the sustained interest of theorists from different disciplines, generating a wide variety of interpretations and undertakings, some of which are divergent in terms of both function and approach. It is therefore the aim of this section to clarify and critique some of the major versions of biopolitics, namely those of Michel Foucault, Giorgio Agamben and Nikolas Rose whose differing conceptualisations are relevant to the present enquiry and to its overall ethico-political concerns. In bringing together these divergent accounts on biopolitics, this section also aims to expose the contradictory and multifaceted aspects of biopolitical modes of governing, aspects that shall be taken up further throughout the course of the remaining chapters and their attendant examples.

Foucault's biopolitics

I begin with the work of Michel Foucault, who is often held as the founder of the concept of biopolitics. In the final part of the first volume of *History of Sexuality*, Foucault juxtaposes the 'bio' and 'politics' as a means of distinguishing between classical politics and modern politics. He writes

> For millennia, man remained what he was for Aristotle: a living being with the additional capacity for political existence; modern man is an animal whose politics calls his existence as a living being into question.
>
> (1979: 143)

Foucault's distinction is marked by a temporal rupture that clearly defines the modern age in terms of the inclusion of 'life' into the political sphere. In addressing this putative historical passage, Foucault focuses on what he sees as profound and defining transformations in the logic of *power*. He argues that one of the major characteristics of the classical age was the sovereign form of power, a power to '*take* life or *let* live' (ibid.: 136). In this sense, sovereign power was exercised at the level of the 'individual' as a means of 'deduction': the ruler's right to seize 'things, time, bodies, and ultimately life itself' (ibid.).

Around the beginning of the seventeenth century, however, the rationale of power started to shift, according to Foucault (ibid.: 139). Deduction was no longer *the* main form of power but only one component of it. And by the middle of the eighteenth century, new mechanisms of power started to emerge with the aim to 'administer' the life of the population as a whole. This meant that the task at hand was no longer merely the discipline of individuals but the management of the *population* in its multiplicity. This meant that the sovereign's old right (to take life or let live) was overridden[9] by the right to '*foster* life or *disallow* it to the point of death' (ibid.: 138). This meant that what became at issue was *life* itself. With this paradigm shift in mind, Foucault goes on to suggest two bipolar models in which the power of life evolved, delineating the beginning of an area of 'biopower'.

The first model of biopower is what Foucault calls the *anatomo-politics* (ibid.: 139). It is designed to seize power over the *human body*, over 'man-as-body', in order to maximise its capacities, increase its usefulness and docility and integrate it into efficient systems. The second model of biopower, on the other hand, is directed at the *species body*, at man-as-species, in order to control and manage the life of the population through statistical norms (e.g. the measurement of birth and mortality rate, longevity, reproduction, fertility and so on). This is the *technique* of biopower that Foucault refers to as *bio-politics* (ibid.). By the nineteenth century, both of these paradigms of power became crucial to the development of capitalism. They provided the mechanisms by which bodies and populations could be managed and rendered more productive and adjustable to economic growth and processes. And for the first time in history, according to Foucault, 'biological existence was reflected in political existence [...]; it was the taking charge of life, more than the threat of death, that gave power its access even to the body' (ibid.: 142–3).

In *Society Must be Defended*, Foucault (2003 [1976]) extends his analysis of biopolitics to include specific domains wherein biopower seems to be at play. The sites, he cites, range from the field of medical care, demographic analysis, natalist policy, urban planning to some more subtle mechanisms such as insurance, individual and collective saving, safety measures and the like. (ibid.: 243–5). These varying domains of biopower are concerned not only with the individual body and its disciplining but also with the population and its regularisation on a massive scale. From the outset, then, it appears that the main task of biopower is to intervene at the level of life in order to improve it, sustain it and increase its chances. However, this is merely one part of the chronicle of biopower: the right to 'make live'. As for the other part, the story is about death: the right to 'let die'.

Herein lies the paradox of biopolitics, the same techniques that are designed to enhance life can be used to expose not only the 'enemies' but also the 'citizens' to the risk of death (ibid.: 254). And this is precisely the point at which the notion of racism comes into play. Racism, for Foucault, is a way of creating caesuras within species-bodies and fragmenting the biological continuum. In doing so, racism allows the (sub)division of the population into manageable groups, some of which are regarded as 'good', while others are perceived as 'inferior' (ibid, 254–5). From here, racism takes up a function that is intimately intertwined with death. It is the function by which 'killing' is made acceptable in order to eliminate biological threats (not only diseases but also the 'bearers' of diseases) and enhance the national stock (through eugenic practices, for instance). For Foucault, what is entailed by the notion of killing is not simply 'murder', but also 'the fact of exposing someone to death, increasing the risk of death for some people, or, quite simply, political death, expulsion, rejection, and so on'[10] (Foucault, 2003 [1976]: 256). At the end, this juxtaposition of racism and biopower culminated into something of an irresolvable puzzle for Foucault. No wonder, then, biopolitics has become such a highly contested field of enquiry.

Agamben's biopolitics

The Foucauldian thesis on biopolitics constitutes one of the many points of departure for Agamben's *Homo Sacer* (1998) in which he seeks to provide an alternative dimension for understanding and conceptualising the notion of biopolitics. Agamben's engagement with Foucault's thesis, however, features primarily as a rejection of the historical break between the classical and the modern paradigm of power, between politics and biopolitics, which seems to characterise much of Foucault's analysis of sovereignty and biopower. Building upon Aristotle's distinction between *zoē*, the *natural* life, and *bios*, the *political* life, Agamben goes on to add a third term to the pair; *bare* life, casting it as the main 'protagonist' of his book (ibid.: 8). As Mills (2004: 46) put is,

> The category of bare life emerges from within this distinction, in that it is neither *bios* nor *zoē*, but rather the politicized form of natural life. Immediately politicized but nevertheless excluded from the polis, bare life is the limit-concept between the *polis* and the *oikos*.

The introduction of the concept of bare life has a considerable impact on Agamben's entire theorisation of biopolitics. It serves as an important analytical tool by which he refutes the historically successionist approach adopted by Foucault. It leads him to claim that life is situated within an originary relation to politics rather than being imported into it through the advance of biopolitics. That the original exclusion of *zoē* from the *polis* is an 'inclusive exclusion' (an *exceptio*). That it is 'almost as if politics were the place in which [...] what had to be politicised were always bare life' (Agamben, 1998: 7). Bare life, as such, represents the original 'nucleus of sovereign power', its 'first content' (ibid.: 83). And in so doing, it makes the Foucauldian successive distinction between sovereign power and biopower rather elusive, according to Agamben.[11]

In this sense, for Agamben, what distinguishes biopolitics from politics is not so much the 'inclusion' of biological life into the political realm, nor is it the rendering of bare life as the principle object of political calculations. Instead, it is the process by which 'exception' is generalised to the point where it becomes the 'norm'. It is the process by which what was once at the margin (bare life) begins to overlap with that which constitutes the centre of the political, and 'exclusion and inclusion, outside and inside, *bios* and *zoē*, right and fact, enter into a zone of irreducible indistinction' (ibid.: 9). Here, Agamben is combining Carl Schmitt's notion of 'the state of exception' in which the sovereign is he who decides the exception with Benjamin's notion of 'the state of emergency' where exception is the rule itself. And within this permanent state of exception, bare life becomes both the 'subject' and the 'object' of the political order whereby the two processes of totalitarianism (where the living being is the *object* of political power) and democracy (where the living being is the *subject* of political power) are concurrently in motion. The peculiar convergence of these two processes is another way of distinguishing modern politics from classical politics and understanding the aporia of modern democracy. That is to say, the placement of 'the freedom and happiness of men [...] in the very place – "bare life" – that marked their subjection'[12] (ibid.: 9).

By invoking the Roman figure of *homo sacer*, Agamben gives flesh to his argument vis-à-vis the way in which bare life is included in the juridical order precisely by means of its exclusion. For *homo sacer* is the one who may be killed without being sacrificed, whose life is exposed and abandoned to violence and death, whose killing is excluded from notions of punishment, execution, condemnation and sacrilege entailed within the realm of law (be it divine or human) (ibid.: 83). Thus, the *sacer* represents an 'ambivalent' character in which the taboo, the impure, the horror and the profane often coincide with the holy and the sacred, confirming the 'double meaning' and the contradictory traits of the *homo sacer* (ibid.: 71–83). So bare life, in this respect, is the life of *homo sacer* (ibid.: 8); the life that has been captured in the sovereign sphere where 'it is permitted to kill without committing homicide and without celebrating a sacrifice' (ibid.: 83). This means that for Agamben, the state of exception is the space *par excellence* where 'all subjects are potentially *homo sacers* [...] abandoned by the law and exposed to violence as a constitutive condition of political existence' (Mills, 2004: 47).

Agamben's disquieting and indeed controversial formulation of biopolitics alongside the notions of exception, bare life and *homo sacer* has received a considerable amount of interest in recent years, and informed a body of literature on disparate topics as well as related ones (including immigration and asylum policy, human rights, euthanasia, biotechnology, surveillance, organisation and management, etc.). For one thing, Agamben's thesis, as Dean (2004: 26) argues, challenges to some extent the Foucauldian bipolar paradigm and displaces the binaries of life/death, sovereign power/biopower, techniques of government/technologies of the self, individualisation/totalisation and so on.[13] In this regard, Agamben manages to, partially at least, resolve the Foucauldian puzzle by elucidating the ambivalent, contingent and fluctuating point at which these binaries coincide and overlap, through the lens of the state of exception (ibid.). Also, in bringing the notion of sovereign power back into focus, Agamben provides the 'exegetical audit' (Dillon, 2005: 43) needed to expose the limitations of the Foucauldian approach towards liberal-democratic politics (Dean, 2004: 26). In so doing, Agamben opens up alternative and discursive ways for attending to the 'violence' intrinsic to some political practices and their emerging and enduring phenomena (manifested, for instance, in the spread of detention centres, refugee camps, border control and anti-terror laws).

Nevertheless, Agamben's thesis on biopolitics has also been the subject of much criticism, mainly for its generalising, over-dramatic and pessimistic tone as well as its lack of historicity. Rabinow and Rose (2003: 8), for instance, contend that Agamben's account is only suited to the twentieth-century absolutisms of the Nazi and Fascist regimes, and fails to recognise the conceptual and historical complexity entailed within Foucault's arguments. More specifically, they are concerned with Agamben's reliance on the figure of *homo sacer*, a figure that is specific to a particular historical era, to explicate the entire working of contemporary biopolitics and substantiate the presence and persistence of sovereign power within current forms of political rationalities and technologies. Mitchell Dean (2004: 27) raises similar concerns in the following way:

> Agamben gives an originary structure to sovereignty in *homo sacer*, which is also materialized in different historical instances and is an actuality in contemporary society. Is there not a problem here of essentialism that seeks a trans-historical form of sovereignty? Is there not a lack of historical sense? Above all, does such a view not miss the rupture identified by Foucault in the eighteenth century?

While such arguments are certainly important in maintaining the needed critical distance vis-à-vis Agamben's account, it should be borne in mind, however, that Agamben (2002) is not a historian, nor does he claim to be so. He is 'a *philosopher* at work' (Bos, 2005: 16) whose *political* engagement places a demand on philosophy to rethink the 'political' itself by taking sides with the refugee, the immigrant, the detainee, the subaltern, the repressed and so on. For he believes that these are the singularities that expose the dark (and all the more constitutive) side of biopolitics and issue a challenge to thinking itself. Undoubtedly, in taking such a stance, Agamben is 'betraying' the Foucauldian thesis (ibid.: 38) at the very same moment he is claiming to be completing it. This 'schema of betrayal' (Nancy, 1991) is, in fact, nothing other than a 'trade-off' between *history* and *politics*. And in the case of Agamben, the trade-off culminates into tearing the concept of biopolitics away from history and placing it (back) into political philosophy (Dillon, 2005: 38). Or again, this schema of betrayal is reminiscent of the all too familiar 'dispute between the thinker who thinks philosophically and the thinker who thinks politically' (ibid.: 42–3).

But while this trade-off can be regarded as a point of weakness in Agamben's approach, it serves, at the same time, as a powerful reminder (in a world struck by oblivion despite its fetishistic tendency to 'consume' history) of the need to probe deeper into some fundamental issues that seem to be, at first glance, excluded from the rationalities of contemporary biopolitics, but which, nevertheless, remain at its very centre.[14] And although the figure of *homo sacer* seems anachronistic at times, this metaphor (here, one may even ask since when a *metaphor* had to be 'historically' constrained), however, describes well the 'unpunishability' aspect of the deaths taking place regularly within or under the gaze of Western democracies.[15]

It remains true though that Agamben's generalising account fails to address the particularities of the cases and figures he uses to illustrate and substantiate his arguments regarding the contemporary logic of biopolitics as well as the *material* reality in which these cases and figures exist. Consequently, this ends up jeopardising the very notions of *singularity* and *alterity*, notions that are important to Agamben himself (see, for instance, his earlier book *The Coming Community*). It remains also true that Agamben's thesis is admittedly incomplete (rather than entirely erroneous or out of context). It is incomplete to the extent it does not cover 'the manifold ways in which the event of biology's biologisation of life continues to mutate, driven by successive changes in the character of the life sciences themselves' (ibid.: 38). It does not address what has become of 'the *bio-* part in biopolitics' (Thacker, 2005). In short, it does not fully attend to the role of techno-scientific developments in articulating and mediating the connection between politics and biology, and transforming the socio-cultural imaginary vis-à-vis 'life itself'. This leads us to another thesis on biopolitics, the one developed by Nikolas Rose.

Rose's biopolitics

Rose's thesis on biopolitics can also be read as an attempt to complete the Foucauldian one. Or more accurately, it is an attempt to 'update' it. In *The Politics of Life Itself* (2001; see also Rose, 2007), Rose engages precisely with that which has been left out in the work of Agamben, that is to say, the role of biotechnology, biomedicine, bioscience and other related disciplines in shaping and configuring the very notion of 'contemporary' biopolitics. In doing so, he provides three main key points of analysis whose arguments hinge on the notions of *risk, molecularisation* and *ethopolitics*.

Rose begins his analysis by pointing out to a set of anxieties currently circulating within the debates on the developments in bioscience and biomedicine. These anxieties are mainly related to concerns over the possibility of resuscitating biological racism and eugenic practices (such as those witnessed during the nineteenth and twentieth centuries) through the technological advances in the life sciences. As a response to these concerns, Rose revisits the biopolitical rationalities developed at the beginning of the twentieth century setting the stage for his later arguments about contemporary biopolitics. Inspired by the work of Foucault, he outlines two forms of state-sponsored biopolitical strategies that were designed to 'maximize the fitness of the population' (ibid.: 3) through purification and elimination. The first was concerned with *hygienic* issues and 'sought to instil habits conductive to physical and moral health' (ibid.), while the second was concerned with *reproduction* and sought to enhance the national stock and 'relieve it of the economic and social burdens of disease and degeneracy' (ibid.). Both of these biopolitical strategies used a combination of 'state-directed' techniques and 'individual' ones (what Foucault calls 'technologies of the self') in order to manage the population en masse. They were underpinned by a mixture of compulsory and voluntary eugenic programmes.

As regards contemporary biopolitics, Rose contends that the political rationalities of our present age differ in many ways. The fitness of the population is no longer framed in terms of a struggle between nations. Instead, it is framed in 'economic' terms (days lost of work because of illness and rise in insurance contribution) or in moral terms (reducing inequalities in health). The management of health is no longer orchestrated by the state alone, but became a distributed agency and a collective responsibility shared between government, individuals and a plethora of non-state bodies. Thus, the state is no longer expected to take charge of the health needs of the entire society, but to merely 'facilitate' and 'enable'. The 'omnipotent' state has now become an 'animator' state. And the distinctions between the normal and the pathological are now organised through actuarial and epidemiological strategies of reducing aggregate levels of risk (ibid.: 5–7).

Risk management, as such, has become central to this 'new' biopolitical order. It has given rise to a number of social insurance strategies and a range of practices that are increasingly geared towards the 'future', and how the future can be calculated, predicted, pre-empted and optimised. Most importantly, for Rose, the contemporary logic of risk thinking indicates a defining *shift* towards 'pastoral' forms of power that are not state administered but diffused across a network of actors and participants (health specialists, professional associations, ethics committees, insurance companies, etc.). Unlike disciplinary power, pastoral power is concerned with 'individual susceptibility' rather than the 'flock as a whole' (ibid.: 9–11). This in turn, according to Rose, has transformed the ways in which the body and life itself are understood and acted upon. Instead of the 'eugenic' body, we are now dealing with the 'genetic' body.

From here emerges the notion of molecularisation, which is the second key argument in Rose's thesis. Molecularisation does not merely indicate a change in the level at which explanations are articulated and artefacts are fabricated. But more so, it is a 'style of thought' about life and how it is 'visualised'. It is 'a reorganisation of the gaze of the life sciences' (ibid.: 13) facilitated through the advances in 'digital' technology. It is 'an irreversible epistemological event [as well as] a significant technical event' (ibid.: 14) in which 'natural life' is no longer the means by which biopolitics can be assessed and judged (ibid.: 17).[16] For life itself has become amenable to processes of shaping and reshaping, to a series of events that can be reengineered, reconstructed and redefined at the molecular level. Hence, biopolitics has become molecular politics.

The combination of risk politics and molecular politics gives rise to a third form of biopolitics: *ethopolitics*. Ethopolitics is Rose's term for today's (arguably) predominant political rationality. And as the name indicates, it is a composite of 'ethics' and 'politics'. It denotes a whole array of relations and practices that are seen to be reshaping how individuals relate to themselves, to the state and to other authorities (public and private). It is a 'normative' modality of thought delineating how life 'should be lived' and generating new ways for making individuals aware of their future risk and able to make informed decisions regarding their health and life in general. As Rose (2001: 18) puts it:

> If discipline individualizes and normalizes, and biopower collectivises and socializes, ethopolitics concerns itself with the self-techniques by which human beings should judge themselves and act upon themselves to make themselves better than they are.

In this sense, individuality has become 'intrinsically somatic' (ibid.).[17] Ethical concerns and practices are now framed not only in terms of personal conduct, but increasingly in terms of the 'corporeal' existence of the self. In this ethopolitical model, individuals are urged to be active, prudent and responsible citizens; able to understand their rights and obligations; and secure their own well-being. Elsewhere, Rose and Novas (2002) take up similar arguments to suggest that a new kind of citizenship is taking place: a *biological citizenship*. They argue that this form of citizenship poses many challenges to traditional concepts of *national* citizenship insofar as it is not taking a 'racialized and nationalised form' nor is it just 'imposed from above' (state). Instead, it is also manifested from below (citizens) and takes as its project the maximisation of *biovalue*[18] rather than racial purity (ibid.: 3). Partaking of what Foucault calls 'the technologies of the self', biological citizenship is an *individualising* process in that it entails a sense of responsiblisation and subjectification (akin to the logic of ethopolitics). It is also *collectivising* insofar as it is giving rise to new notions of solidarity and new forms of sociality[19] based around people's experiences of dealing with diseases, heath providers, insurers and so on (ibid.: 6).

In a way, Rose's analysis of contemporary biopolitics might be closer (than Agamben's) to providing the sequel for the Foucauldian saga of biopolitics. It follows on precisely from what Foucault was set about to do: exploring the *historical* mutations of life through the developments in the techniques and technologies of biopower – with medicine being a primary ground of analysis. However, and despite the merit and insight of Rose's analysis, his exclusive focus on the life sciences[20] and on the autonomous, responsible and informed individual elides some of the issues that are equally salient and pertinent to understanding contemporary biopolitics. Ultimately, this makes Rose's account also inescapably incomplete. For if the protagonists figuring in Agamben's work are *homo sacers*, refugees, immigrants, tramps, inmates of detention centres and concentration camps and so on, in Rose's texts the principal protagonist remains the (affluent) right-bearing citizen (see also Braun, 2007). And if in Agamben's analysis, there is a tendency to conflate biopower with sovereign power, in Rose's approach, there is a tendency to conflate biopower with pastoral power.

In this sense, one of the marked problems in Rose's take on biopolitics is that his analysis does not engage with *other* 'spaces' and 'bodies', which are increasingly becoming not only the site but the *product* itself of biopolitical interventions. Despite this, however, Rose's account presents us with a useful set of analytical tools under the umbrella terms of risk politics, molecular politics and ethopolitics. But these concepts are also crying out for *more* exploration and experimentation than Rose's version could bestow. Braun (2007), for instance, links up Rose's notion of biopolitics to the field of 'geopolitics', precisely in terms of 'biosecurity', in order to reveal how the government of life goes hand in hand with 'the global extension of forms of sovereign power whose purpose is to pre-empt certain biological futures in favour of others' (ibid.: 6). In doing so, Braun foregrounds the theoretical and empirical imperative of bringing conflictual accounts (governmentality and sovereignty) together in order to understand how they 'relate' to each other.

It is from a similar vantage point that I regard Agamben's and Rose's theses on biopolitics as not being mutually exclusive but complementary to each other. They are not necessarily antithetical but symptomatic of the complexity and hybridity inherent to the notion of biopolitics and its practices. So by juxtaposing their divergent mindsets in this book (without, of course, overlooking their specificities and distinctive arguments), I am aiming to surmount the epistemological rupture between the two as well as the overall 'metanarrativisation' of biopolitics. This is in order to derive a balanced (albeit contradictory) framework of analysis that is able to attend to those complexities and multiplicities intrinsic to the biopolitics of biometrics and identity systems.

Conclusion

In situating biometrics in its historical context, this chapter has demonstrated the relation of this technology to its predecessors in order to challenge the label of newness that is often stapled on it and to draw attention to the fact that the body has for so long been the subject of control, measurement, classification and surveillance. The digitisation aspect of biometric has certainly intensified such processes and opened up the body to further dynamics of power and control. Some of these will be explored in detail throughout the book.

The chapter then moved on to consider the concept of biopolitics, which forms the key theoretical base of this project. From Foucault's initial conceptualisations to Agamben's reworking of the concept and ending with Rose's approach to biopolitics, this section of the chapter provided a critical discussion on these varying accounts, explaining their specificities and differences. Each of these accounts on biopolitics brings relevant and useful insights into the analysis of biometrics and identity systems. For instance, Agamben's reformulation of biopolitics in relation to the 'state of exception' touches, and penetratingly so, upon a number of matters pertaining to the politics of identification in which the body is included in the (bio)political strategies and interventions, and made accessible to technologies of control. It offers an incisive way for understanding the mechanisms by which what was once confined to 'exceptional' spaces and practices is now in the process of becoming a permanent rule by spilling over to the 'biopolitical body of humanity' as a whole (Agamben, 1998: 9). In this respect, the logic of the state of exception may help us understand the 'functional creep' of biometric identity systems and reveal their close interplay with sovereign forms of power (see Chapter 2).

While, on the other hand, Rose's concepts of risk politics, molecular politics and ethopolitics incite us to explore the emerging technologies of biopower in a way that cannot possibly be captured through Agamben's figures of *homo sacer* and the camp. Afar from biomedicine and bioscience, these concepts have also the potential to establish a useful and extendable framework for analysing the interface between other biotechnologies (biometrics being a biotechnology) and contemporary biopolitics. They can tellingly be deployed to articulate the role of biometrics in mediating and transforming notions of the body, identity and identification through mechanisms of biologisation, informatisation and digitisation. Moreover, the concept of biological citizenship is equally relevant to the study of biometric identity systems insofar as it subsumes 'all those citizenship projects that have linked their conceptions of citizens to beliefs about the biological existence of human beings' (Rose and Novas, 2002: 2). In doing so, it can expose some of the features of the changing relationship between the state and *its* citizens within the era of molecular biopolitics. While this concept of biological citizenship is often used to discuss the formation of active (neo)liberal subjects endowed with rights, choice and responsibility (as is the case in Rose and Novas' work) (see Chapter 4), it can also be used as a means for uncovering the potential role of biometric identity systems in enacting contemporary surveillance practices at a larger social scale, enhancing the state's 'embrace of individuals' (Torpey, 2000: 166) and, ultimately, turning society into a *biomass* rather than a political community.

2
Homo Carded: Exception and Identity Systems

The issue of 'function creep' is one of the most recurring concerns in the debates surrounding the implications of biometric technology and ID cards systems. Underlying these concerns is the fear that the use of biometrics may overflow beyond its originally intended purposes, especially where the concept of 'interoperability'[1] and technologies of 'networked' databases are involved (van der Ploeg, 2005b: 13). To be sure, this concern over the function creep of biometric technology is all too often articulated in relation to the increase in surveillance practices, and more specifically, in relation to the issue of privacy and the problem of data misuse. Thus the debate over function creep continues to be largely confined to what *technology* can and cannot do, and what possible uses and scenarios may ensue. As the following statements indicate:

> 'Function creep' is an important concern, i.e. that technology and processes introduced for one purpose will be extended to other purposes which were not discussed or agreed upon at the time of their implementation.
>
> (European Commission, 2005a: 16)

> Just as function creep implies that biometrics will gradually (and innocently) grow to be used by zealous, well-meaning bureaucrats in numerous, creative ways in multiple fora, function creep will also enable the Government to reduce further over time the citizenry's reasonable expectations of their privacy.
>
> (Woodward, 1998: 12)

> The first applications of biometric technologies are for very limited, clearly specific and, for the most part, sensible purposes.... But the

greatest danger would be the expansion of such use for well-meaning purposes to other that went beyond the original purposes and failed to address the limitations of the original collection activity.

(Cavoukian in Aus, 2003: 34)

The notion of function creep is nothing new; the same process happened with the ID card issued during World War II when there were originally three purposes for the card (national service, security and rationing); eleven years later thirty nine government agencies made use of the records for a variety of services.

(LSE, 2005: 149)

While such articulations have their own merit in highlighting the functional stakes of biometrics, I argue that more efforts are yet to be made to locate the function creep of biometric technology within a wider *political* frame of analysis so as to attend more closely to how the 'normative vocabulary' of function creep 'translates in the *lives* of people' (van der Ploeg, 2005b: 13, my italics). Such a task would necessitate a broadening up of this very vocabulary in such a way that what is underscored as being at stake is not only mere data, but the embodied existence itself (van der Ploeg, 2003a: 71–2). So how are we, then, to proceed? I propose the notion of *exception* as a possible avenue for analysing and scrutinising the ontological and procedural overflow of biometrics.

When biometric identity systems were somewhat an exception

The initial social and political use of biometric technology was limited to exceptional spaces and extreme cases, spaces such as detention centres and cases such as crime investigations. In order to understand the function creep of biometric technology and how it is increasingly spilling over from such exceptional spaces and practices into the 'biopolitical body of humanity' (Agamben, 1998: 9), it is first necessary to explore some of those initial examples whereby biometric technology started to be (and is still being) deployed in ways that reinforce the logic of exception and epitomise the working of its biopolitics. I start the discussion by looking at the introduction of biometrics as part of the Eurodac project whose *raison d'être* is to control 'illegal' immigration and border crossing by 'asylum seekers' in Europe. (Here I focus on the EU example. For a similar example from the US, see the case of the Ident system which was implemented in 1997 along the borders between the US, Mexico and Canada. This system contains two databases the 'Lookout' and the 'Recidivist' designed for the comparison of fingerprints of asylum seekers and 'illegal' migrants. See also van der Ploeg's (2005a) discussion.) I will then move on to examine, at length, the issue of borders and the technologies deployed and envisioned for their securitisation, with particular reference to the UK White Paper *Secure Borders, Safe Haven* as it constitutes one of the initial governmental proposals and crucial steps towards the current developments and changes in the UK's immigration and citizenship policy landscape. The third example will look at the introduction of biometric Application Registration Cards (ARCs) for asylum seekers in the UK.[2]

Eurodac

The Eurodac project is a European Union initiative aimed at facilitating the implementation of the 1990 Dublin Convention concerning the criteria and mechanisms for determining the Member State responsible for the examination of an asylum application (European Union, 2006). The Convention was established in the context of developing a common and *harmonised* European asylum system – in other words, 'the "Communitization" of asylum policies' (Aus, 2003: 8). It is governed by the 'authorisation principle' (Hurwitz, 1999: 648), which lays down the rule that the State of first entry would be the one and the *only* Member State who has total jurisdiction in and responsibility for the asylum application. In this way, if the application is rejected by one Member State, the asylum seeker will not then be able to apply in any of the other Member States (Koslowski, 2003: 9). As such, the Convention has two major goals. First, it is designed to combat what is referred to as 'asylum shopping' (ibid.; van der Ploeg, 1999a: 298) by preventing the lodging of multiple asylum applications by the same person in several Member States. Second, it is aimed at putting an end to 'orbit situations' by obliging the responsible State to process the asylum application rather than passing it onto another State (Hurwitz, 1999: 649).[3] The Eurodac project was proposed in 1997 and went live in 2003 as a response to the problem of determining applicants' prior stay in other Member States. In this respect, establishing reliable and effective techniques for *identifying* and *verifying* the identity of each asylum seeker is considered as a crucial element for the feasibility, success and optimisation of the Eurodac initiative.

Underlying the Eurodac project is a supranational cybernetic network, an EU-wide database that is 'the first common Automated Fingerprint Identification System (AFIS) within the European Union' (European

Commission, 2005b). It contains the digital fingerprints of every person over the age of 14 who is claiming asylum in one of the EU countries (ibid.). Prior to assigning any given asylum application to a case worker, the applicant's fingerprints are taken and matched against other digitised fingerprints in the database of the Central Unit that is responsible for the storage and matching of new fingerprints against those already stored. The aim of this biometric process is to establish whether an applicant has already tried claiming asylum at another border crossing. If a match is found between the applicant's fingerprints and others that are already stored in the central database, she or he will then be subject to deportation to the country of the first application if not to the third country of origin.

The practice of digital fingerprinting was soon *extended* through a 'separate Protocol' to cover the issue of 'illegal immigration' as well, despite the objection of some Member States, such as France and Luxemburg, to the 'unconformable amalgam' of the issues of immigration and asylum (van der Ploeg, 1999a: 299). The Protocol itself was later included in the main Eurodac Regulation, making it possible for Members States to record, transmit and match the fingerprints of those regarded as 'irregular' border-crossers and those found on European territory 'illegally' (i.e. without the necessary residency or identity papers) (Brouwer, 2002: 235). This substantive move was considered by some European politicians as a necessary step to 'curb the entry into the EU of *illegal refugees*' (EU Council, cited in Brouwer, 2002: 235, my emphasis) and deal with the 'Influx of migrants from Iraq and the neighbouring region' (ibid.). The former German Interior Minister, Manfred Kanther, and his Secretary of State, Kurt Schelter, were the major instigators of this shift and strong proponents for exploiting the 'added value' of the Eurodac system (Aus, 2003: 12–3). Kanther's successor, Otto Schily, later expressed his political ambition to make Eurodac available for 'general police purposes' (ibid.: 20).

From here, it can be noticed how the functions of Eurodac system began to gradually expand beyond their initial intended target, and in ways that collide with the original Dublin Convention.[4] As observed by Aus (2003: 12), the 'extension of the Eurodac database to irregular border-crossing and illegal residence is a striking incident of policy "spill-over" and provides one of the theoretical "puzzles" of this seemingly asylum-centered project'. Striking too is how the 'biometric control of asylum seekers and "illegal" immigrants in the European Community [...] is very similar to police practices previously employed on nation-state level vis-à-vis ordinary *criminals*' (ibid.: 23–4).

The practice of fingerprinting, for instance, is a major hallmark of this *criminalisation*. And it is important to remember, at this point, that in the context of 'criminal law', the taking and recording of fingerprints is (or at least was) only allowed in serious criminal offences where suspects may be taken in custody or detained on remand (van der Ploeg, 1999a: 300).[5] However, in the context of immigration and asylum, these standards and the 'conditions of proportionality' surrounding the overall practice of fingerprinting are loosened quite significantly (ibid.) to the extent that the fingerprinting of asylum seekers and 'illegal' immigrants is now an unquestionable 'common practice' in most European countries (Brouwer, 2002: 243) – the *only* criterion is one's (unfortunate) identity, one's 'irregular existence', so to say: being 'born into the wrong kind of [...] class or drafted by the wrong kind of government' (Arendt, 1966: 294).

In fact, in Holland, for example, the fingerprints of asylum seekers were, for many years, stored together with the fingerprints of 'criminal suspects' on the same database ('HAVANK'), with no legal or technical separation (Brouwer, 2002: 243). This allowed the use of asylum seekers' fingerprints for criminal investigations (and as van der Ploeg (2003a: 60) argues, just the act of *retrieving* a stored fingerprint may indeed generate a suspect). It was not until September 2001 (when the *Dutch Protection Act* came into force following the EC Directive 95/46) that the Dutch Minister of Justice promised the implementation of technical procedures by which data recorded on asylum seekers cannot be used for the purpose of criminal investigations – although both categories of fingerprints will remain stored on the same database (Brouwer, 2002: 243).

Ultimately, this modality of biometric control ends up contributing to the conflation of foreigners and refugees with criminality and illegality[6] (van der Ploeg, 1999a: 300), especially when combined with the escalating discourses of fear and otherness whereby criminality and asylum are placed side by side: 'We will explore what more we can do, as other countries have done, to stop serious criminals abusing our asylum system' (Home Office, 2002a: 68). This linguistic coupling of asylum and crime generates confusion between the two, placing immigrants and asylum seekers under sustained suspicion and fostering a siege mentality (see also Rajaram and Grundy-Warr, 2004: 41–2). It thereby creates the opportunity to 'govern through crime' (Simon in Walters, 2004: 247), to manage global circulation through processes and techniques of criminalisation and illegalisation (Nyers, 2003; Walters, 2004: 247; Yuval-Davis et al., 2005: 516) as is the case with Eurodac system.

It is thus not so surprising that the criminalisation of asylum seekers and immigrants has become one of 'the globally relevant features of the contemporary discourse' (Diken, 2004: 88). It is the *performative* act by which undesirable and uninvited bodies are (compulsorily) made accessible to biopolitical technologies of control, such as those of biometrics, in order to establish the boundary between the 'genuine' and the 'bogus', between the 'legitimate' and the 'illegitimate'. The politics of immigration is, as such, a way of distinguishing the *polis* from 'what does not "properly" belong to it' (Zylinska, 2004: 526).

Before proceeding further, it is worth saying a word or two about the politics of immigration. This will help us understand the double logic embedded within asylum and immigration policy, and throw some light upon that which constitutes the condition of possibility for enacting 'illiberal practices' (including compulsory fingerprinting and other biometric modes of control) within 'liberal regimes' (Bigo and Tsoukala, 2006). This will also pave the way for our discussion on the relationship between the use of biometrics for immigration and asylum control, and the notion of exception, while giving a touch of 'particularity' to the generality of the latter.

Biopolitics of immigration

In February 2002, the British government published its White Paper *Secure Borders, Safe Haven* outlining a set of mechanisms for tackling 'bogus asylum applications' and 'illegal immigration', and reinforcing 'community cohesion' within the British society. The title itself is indicative of the double aim and challenge of contemporary politics, namely that of security (of what is constructed as 'legitimate' inside) and protection (of 'genuinely' endangered outside). The politics of immigration stands, in fact, at the crossroad between the politics of borders and the politics of protection. In doing so, it curiously straddles both the logic of inclusion and that of exclusion, and concomitantly partakes of practices of (human) rights and those of (symbolic and physical) violence. It is the crucible of contradictory, asymmetrical and competing functions, discourses and meanings. To elucidate what is meant by this statement, let us dwell for a moment on the notion of 'border' as well as that of 'protection', and examine the close interplay and the burning tension between the two.

Recently, there have emerged numerous attempts to rethink the 'question' of borders beyond the traditional concept of border as a 'line' that simply demarcates the boundaries between the inside and the outside. Etienne Balibar's (2002) *Politics and the Other Scene* remains perhaps as one of the most incisive, lucid and penetrating analyses of what has become of the notion of borders. Here, Balibar argues that borders no longer act as mere territorial dividers that separate the spatial particularity of one state from another. But more so, borders are becoming symptomatic of 'the establishment of a world apartheid, or a dual regime for the circulation of individuals' (ibid.: 82).[7] Recursively, this duality of circulation is also what endows borders with their heterogeneous functionality and multi-layered meaning. As such, Balibar identifies 'three major aspects of the equivocal characters of borders' (ibid.: 78). First, he begins by what he terms the *overdetermination*[8] of borders, a feature that is intrinsic to 'the *world-configuring* function they perform' (ibid.: 79). So in addition to its symbolic function vis-à-vis the state of self-determination, the border also demarcates the politico-anthropological, social, cultural, linguistic and economic differences, and marks the overlapping of various historical moments.

Indeed, this overdetermining feature finds resonance throughout the entire discourse of the UK *Secure Borders, Safe Haven*, especially in terms of how the notion of secure borders is proposed as a major vehicle for enhancing community and race relations, and preserving a sense of 'Britishness'. Worth bearing in mind as well that the White Paper was released in the wake of specific 'events', including September 11 as well as the civil disturbances in the Northern cities of Bradford, Burnley and Oldham where violent standoffs between the police, British Asian Muslim youth and the far right took place in 2001. Nevertheless, these events (important as they are in providing the driving force for introducing various security measures and immigration control mechanisms) are only small instances compared to a long and complicated history of immigration in Britain. As Walters (2004: 239) explains:

> the White Paper is [...] still within the racialized logic that has marked Britain's approach to immigration policy since the 1960s. This is an approach based on the fear that 'uncontrolled' immigration will inevitably result in 'racial tension'.

Therefore, the border, or at least *how* it is conceived and mobilised, is *overdetermined* inasmuch as it is both an expression and an outcome of myriad socio-cultural, geopolitical and historical factors and operations that continuously criss-cross and overlap. The mounting concern over its security is part and parcel of the ongoing efforts to find panacea for the 'imagined' ills, thought to be brought about by the increase in global movement, especially the movement of the 'unwanted'.

The second aspect relates to the *polysemic* nature of the border (Balibar, 2002: 81); polysemic in that it creates different phenomenological experiences for different people. For instance, we know well, all too well, that crossing the borders of Europe with a European passport is not the same as crossing them with an African passport.[9] So while some are endowed with the right to 'smooth passage', others are enduring an 'excess of bordering':

> The challenge here is to allow those who qualify for entry to pass through the controls as quickly as possible, maximising the time spent on identifying those who try to enter clandestinely.
>
> (Home Office, 2002a: 17)

> We will develop processes which allow entry to the UK to be automated by using biometrics technology, such as iris or facial recognition or fingerprints [...] not only to detect and deter clandestine entrants, but also to increase further the effectiveness of the control and the speed at which certain passengers will be able to pass through on their arrival in the UK.
>
> (ibid.: 95)

In this respect, the notion of border acquires different meanings and yields different material and psychological experiences that have almost nothing in common except a name: *the* border (Balibar, 2002: 81). For that matter, Balibar argues that borders are designed to establish 'an international class differentiation' whereby the rich asserts a *surplus* of rights (ibid.: 83) through, for instance, her 'worthy' passport/ID card, whereas the poor continues to exercise, what we may imagine as, the Sisyphean activity of circulating upwards and downwards until the border becomes a place where she resides, or until she becomes the 'border itself'. As vividly and poignantly illustrated by Raj's (2005: 9) depiction which warrants quoting at some length:

> The border for some – those who travel through the Chunnel by Eurostar daily from London to Paris, or the travel writer, who on a whim decides to attempt a reverse crossing from Europe into Africa and manages to do so on 'her own' – is an indication of a 'surplus of rights'. For others, however – those who live in Sangatte and those who roam the port at dusk seeking access to a ferry or a freight train, for the Moroccan children who are dumped at the border daily or 'cleansed' out of their sleeping hideouts at night by

municipal sanitation trucks, but attempt to climb the fence back into Ceuta at night, or the Africans at Calamocarro who staged a riot (only to find themselves deported) when it turned out that all Kurds in Ceuta were granted political asylum – the border becomes 'an extraordinarily viscous spatio-temporal zone, almost a home' as they repeatedly encounter and are regulated by it (Balibar 2002: 83; *Guardian* 15.vi.2001, 17.viii.2001; *Atlantic Monthly* Jan. 2000). So, when the sub-prefect of Calais compares the town port to a piece of Gruyère, or when a British travel writer tells us how easily she and her husband negotiated the mêlée of a crossing into Africa 'on their own' on an intrepid adventure, we are witnessing a double gesture at the level of ideological discursive production (*with material effect!*) (*Guardian* 7.viii.1999; 11.xii.1999). The first gesture is one of stigma- tization and dehumanization [...]: the residents of Sangatte become mice, or worse, rats eating through cheese, and the Moroccans and Africans climbing over fences equipped with movement sensors and bathed in halogen light are no longer people for they do not possess the right papers.

(*Atlantic Monthly* Jan. 2000; *BBC* 19.iv.2004)

For these 'people', the border becomes a dangerous zone behind which the 'life-and-death question' (Balibar, 2002: 77) lurks relentlessly. It becomes a zone of *negotiation* whereby identities, existences, modes of presence are contested continuously (ibid.: 90). Ultimately, the border becomes a 'home'. A primary example is to be found in the proliferating phenomenon of transit zones and detention centres in which 'await- ing' populations are placed '(sometimes for several years, sometimes in a periodically repeated fashion)' (ibid.), and sometimes even fatally. This is indicative of the polysemic and paradoxical function of borders whereby the points of transit and transition are spreading, as Marc Augé (in Fuller, 2003b) observes, 'under luxurious or inhuman conditions'. That is to say, either in the form of five stars hotel chains or in the form of refugee camps; either in the form of VIP lounges or in the form of detention centres.

To be sure, not only do borders work differently and unequally for different people according to their status and where they come from, but they also do not work in the same way for 'things' and 'people'. This is in the sense that while the free movement of capital, commodi- ties, information and so on is encouraged as it sustains the doctrine of free market and perpetuates global capitalism, the circulation and flow of people, on the other hand, is continuously and vigorously filtered, hindered and blocked. At another layer, this also raises what Balibar (2002: 91) calls 'the empirico-transcendental question of *luggage*', in others words, 'the question of whether people transport, send, and receive things, or whether things transport, send, and receive people' (ibid.). Such a question immediately triggers the need to reflect on the fact that, with regard to borders, some people are being treated as things, whether in terms of the various technologies of scanning they are subjected to at the border,[10] or whether in terms of the circuits of trafficking they have to go through before they reach the 'deadly' border.[11] This blurring of boundaries between, what Fuller (2003b) calls, 'ontological textures' amounts to the 'unpeopling' of people, rendering them, to use Perera's (2002) powerful phraseology, as the 'stuff of contraband: traffic, illegals, human cargo. Non-people':

> We are investing heavily in new technology designed to detect persons concealed in vehicles or containers destined for the UK. Such technology includes:
>
> - X/gamma ray scanners: purchased to be placed initially at Dover and Coquelles. We then intend to extend rapidly their availability at other points of entry. The Immigration Service is already sharing scanners owned by Customs & Excise.
> - Heartbeat sensors: trials held at Dover and Coquelles of sensors which can detect, by its movement, the heartbeat of a person concealed within a stationary vehicle.
> - Millimetric wave imaging equipment: Eurotunnel have been conducting trials with millimetric wave imaging equipment which they are using to good effect in Coquelles. We are following the trials with great interest and monitoring the effectiveness of the equipment which senses radiation emitted from within a vehicle.
>
> (Home Office, 2002a: 96)

The third aspect refers to the *heterogeneity* and *ubiquity* of borders (Balibar, 2002: 84). Put simply, borders are no longer to be found *only* at the border. They are no longer constituted around the 'physical' alone. Instead, borders are now infinitely and invisibly *actualised* within and immanent to mundane processes of 'internal' administration and bureaucratic organisation. As such, borders are everywhere, or at least, 'wherever selective controls are to be found' (ibid.: 34). This can range from some ostensible practices such as stop-checks[12] 'inside' the territory or at its shifting periphery, to some more subtle mechanisms such as access to public health services and social benefits, applying for National Insurance Number, applying for a bank account, smart ID cards and so on. For these activities can also function as an *inner border* and a filter of legitimacy.

From here transpires a dialectical movement wherein the border (as we *know* it) is both 'multiplied' and 'reduced', 'thinned out' and 'doubled' so much so that the 'quantitative relation between "border" and "territory" is being inverted' (ibid.: 92). In other words, it is the 'becoming-country' of a border: the border as a *zone* rather than a *boundary*, the border as a *home* rather than a *line*. This means that borders are no longer (at) the edge of the political. Instead, they are now 'within the space of the political itself' (ibid.) and as such at the root of the very 'ideality' that continues to govern identity, citizenship, rights, obligations and so on. Thus, the border emerges all the more as an institution as well as 'a condition of possibility of a whole host of institutions' (ibid.: 84) whose management often coincides with the management of identity and with it the management of body through (biometric) technology. What issues from the triangulation of these formulations (border-as-zone, border-as-home and border-as-institution) is a problematisation of the entire onto-phenomenological experience of border-crossing in terms of politics and life/death, which renders the 'political' question of immigration a quintessentially 'biopolitical' one.

> You can be a citizen or you can be stateless, but it is difficult to imagine *being* a border.
>
> (André Green, in Balibar, 2002: 88)

However, in order to understand what is entailed in the 'biopolitics of immigration', one has to *at least* 'imagine' what it is like to be a border,[13] what it is like to live as a threshold being whose existence is neither this nor that, whose presence is neither here nor there. For this is increasingly becoming the 'reality' of thousands of people to the extent that we can now comfortably, or rather *un*comfortably, assert that 'being a border' is an 'ontological state' *par excellence*. And ontologically speaking, to be a border is to live a life that is a 'waiting-to-live, a non-life' (ibid.: 83). The biopolitics of immigration is precisely the management of that *waiting-to-live*: the waiting-to-live of those who are forcibly placed in overcrowded detention centres where they are subjected to the gaze of constant surveillance and body scanning; those whose asylum or immigration applications are being processed in the

Home Office (a process that may take several years, lengthening the state of 'limbo' in which many asylum seekers and immigrants find themselves); and those who, for the lack of space in detention centres, are released or rather '*abandoned*' on the streets without being granted permission to work or access to support.

The biopolitics of immigration is also the management of *non-life*: the non-life of those whose cases failed and they are, therefore, subject to compulsory deportation. The non-life of '*les sans-papiers*' who are made to endure a constant state of anxiety and fear for not having residence or work permits, and they are hence forced to succumb to cheap labour and harsh working conditions. The non-life of what Peter Nyers (2003: 1070) calls *deportspora*: an abject diaspora created through the multiplying 'transnational corridors of expulsion'.

The biopolitics of immigration is also the (mis)management of *death*: the death of thousand of refugees and so-called 'clandestine' migrants drowned in the sea (for instance, in the Strait of Gibraltar which is argued to be turning into the world's largest mass grave), asphyxiated in trucks (as was the fate of 58 Chinese immigrants who died in 2000 inside an airtight truck at the port of Dover), crushed under trains (the case of the Chunnel Tunnel) and killed in deserts (the US–Mexican border, for instance). In short, it is the management of the death of the *superfluous* (Balibar, 2002: 142), of what Bertrand Ogilvie disturbingly calls '*l'homme jetable*': the disposable human being that no one wants or needs.[14]

But there is also another aspect to the biopolitics of immigration, one that can hardly be overlooked. It is the management of *life*: the life of the (almost) belonging group – quasi-citizens, temporary-citizens, waiting-to-become-citizens, potential-citizens and so on. In short, all those who can demonstrate their 'legitimacy' and 'utility'. In fact, when it comes to contemporary immigration policy, legitimacy and utility are inextricably linked with no chance of separation. And their link is one of the mechanisms of 'sifting through the desirable and the undesirable forms of migration and people [...in order to] extract maximum economic benefits from global processes' (Yuval-Davis et al., 2005: 517). Markedly, in *Secure Borders, Safe Haven*, the government recognises, 'Migration brings huge benefits: increased skills, enhanced levels of economic activity, cultural diversity and global links' (Home Office, 2002a: 9). Therefore, certain forms of immigration, such as skilled labour are encouraged and being managed through a variety of various schemes such as the 'Highly Skilled Migrant Programme',[15] 'Points-Based Work Permits' and

'Sector Based Scheme'. These programmes are an instantiation of 'economic elitism' (Cohen, 2003: 72) insofar as they act as a means of 'distributing [/reducing] the living in [/to] the domain of value and utility' (Foucault, 1979: 144) while establishing and maintaining the boundaries between those who can contribute 'more to the public purse' (Spencer, in Cohen, 2003: 73) and those who have 'little or nothing to contribute' (Cohen, 2003: 73). In this regard, immigration policy is increasingly becoming akin to 'the running of a business' (Walters, 2004: 244), a notion that comes to the fore most notably in the type of idiom adopted throughout the White Paper:

> The UK allows the entry of some highly skilled migrants through our existing *entrepreneurial* routes, such as the Innovator Scheme and our employment-driven work permit system. We have now launched the Highly Skilled Migrant Programme. This represents a further step in developing our immigration system to *maximise* the *benefits* to the UK of high *human capital* individuals, who have the qualifications and skills required by UK businesses to *compete* in the *global marketplace*.
>
> (Home Office, 2002a: 42, my italics)

And since nowadays, every business has to rely one way or another upon (information) technology to survive the economic ruthlessness of global market and maximise its 'competitive edge', immigration policy is also relying on its 'heavy investment' in myriad technological solutions to *organise* the running of its business and ensure that only the 'right kind of others' (i.e. the skilled and the needed; the 'neo-proletariat') are allowed to enter the 'business premises'. This is being supplemented by a process of 'outsourcing' whereby control over the flux of movement is conducted at a distance – for instance, through stringent visa systems and in consulates located in the third countries[16] – in order to keep the poorest foreigners and potential asylum seekers away as far as possible from the EU frontiers (Bigo, 2005a: 6; Yuval-Davis et al., 2005: 518). In this way, biometric technology is adopted as a means of facilitating and automating the triage of identities through identification and authentication. It works just like 'anti-virus software'[17] designed to block the *infiltration* of the undesirable (human cargo) on the one hand, while allowing the crossing of the desirable (human capital) on the other. Biometric technology can thus be regarded as linking between the different aspects of the biopolitics of immigration (the management of waiting-to-live, non-life, death and life) in a way that contributes to the twofold move embedded within the latter whereby the border is being simultaneously, selectively and unevenly closed and opened up.

The business-like modality of thought that underpins and governs immigration policy has also been gradually creeping into asylum policy raising concern about the linking of asylum issues to economic issues, 'what some have called cherry picking of refugees on grounds of skills and potential for assimilation rather than need for protection' (Yuval-Davis et al., 2005: 518). And in addition to that, there are other reasons why one needs to be sceptical about any taken-for-granted understanding of or absolute faith in state protection and thereby approach the second half of the White Paper, that is *Safe Haven* with some caution.

> The ultimate aims of the asylum system are to determine who is and is not in need of protection. Providing a safe haven and integrating quickly into UK society those who are in need of such protection and to remove quickly those who are not.
>
> (Home Office, 2002a: 14)

> Whenever a state ponders whether or not to grant asylum to an individual, it is making an intervention in the politics of protection. This is a significant political issue because the capacity to decide upon matters of inclusion and exclusion is a key element of sovereign power.
>
> (Nyers, 2003: 1071)

In one sense, it can be argued that the state's ability to decide on who will and will not be provided with protection turns the domain of asylum into a site of *biopower*: the power to 'make live' (those who are granted asylum) and 'let die' (those who are deemed undeserving or not genuinely in need of protection). Of course, in the White Paper, what remains explicit is merely the first function of biopower; the making-live, articulated through the catchphrase of 'safe haven'. As for the letting-die function of biopower, it is concealed behind the dense discourse of legitimacy and security. Yet, one can hardly fail to hear its echo reverberating each time the threat with removal and deportation is invoked. In fact, the promise of protection and the threat with expulsion work in tandem throughout the White Paper. And while the British government is priding itself on its 'humanitarian asylum process' (Home Office, 2002a: 52), which makes Britain a safe haven for those fleeing persecution, it is also 'establishing an inter-Departmental group whose task [...is] to develop further a more coherent Government-wide strategy to support the "no safe haven" policy' (ibid.: 104). These paradoxical statements and practices reveal the concomitance of the two functionalities of biopower and the inherent ambivalence of asylum policy.

In another sense (but one that is not all too separate from the first one), the fact that the state has a tendency to 'monopolise' the decision on who will and will not be granted protection and determine alone the terms and conditions for such protection, turns the issue of asylum into a site of *sovereignty* as well. As Nyers (2003: 1071) argues, this monopoly is not only a 'humanitarian determination' but 'a crucial source of legitimacy for sovereign power' and an important moment for various 'national and international (re)foundings' (ibid.: 1090). So if the discourse of protection might give us a reason to think that the so-called 'humanitarian asylum process' places a limit upon the working of nation-state sovereignty, it also gives us a reason to believe that a different mode of sovereignty is constantly at work. This mode of sovereignty is nothing other than 'bio-sovereignty: a form of sovereignty operating according to the logic of the exception rather than law, applied to material life rather than juridical life' (Caldwell, 2004: 3).[18] For each time the government 'intervenes' in the politics of protection, it 'raises as a question the status of life, and calls for a sovereign decision on life' (ibid.: 11). Each time it acts to decide on the fate of asylum seekers or those who lost every possible status except 'the abstract nakedness of being human and nothing but human' (Arendt, 1966: 297), it produces along the way a 'detritus humanity' (Rajaram and Grundy-Warr, 2004: 35); a humanity that is no longer human and one that is made amenable to various excessive *organisational* arrangements, including those of detention, deportation, compulsory fingerprinting and so on. Hannah Arendt (1966: 297) reminds us that '[o]nly with a completely organized humanity could the loss of home and political status become identical with expulsion from humanity altogether'.

Therefore, the discourse of safe haven should not deceive one, nor should one be surprised that the state's humanitarian organisation[19] of and intervention in the sphere of protection is coexistent with forms of violence that capture the naked life of asylum seekers and irregular migrants in the web of its bio-sovereign power. At times, it is (still) in the name of protection that violence is exercised. At times, it is the logic of humanitarianism itself that justifies exceptional practices imbued with sovereign intentions and impulses. As Žižek (2002: 91) argues, 'the privileged object of the humanitarian biopolitics [...] is deprived of his full humanity through the very patronising way of being taken care of'. He goes on to add that refugee camps and the delivery of humanitarian aid 'are the two faces, "human" and "inhuman", of the same socio-logical formal matrix' (ibid.).[20] The following statements are indeed a case in point:

> We propose to extend the existing power of detainee escorts to search detained persons to allow their entry to private premises to conduct such searches. Searching people being taken into detention is necessary to ensure the safety and security of the detainees themselves as well as of those escorting them.
>
> (Home Office, 2002a: 68)

> We have expanded the number of immigration detention places from about 900 in 1997 to just under 2,800 by the end of 2001. The new Removal Centres at Harmondsworth, Yarl's Wood and Dungavel which opened during 2001 accounted for 1,500 of these additional places. We have decided to increase detention capacity by a further 40%, to 4,000 places, in order to facilitate an increased rate of removals of failed asylum seekers and others with no basis of stay in the UK. Work to identify suitable sites is underway and we expect to have all the additional places in operation by Spring 2003.
>
> (ibid.: 66)

> It will always remain necessary to hold small numbers of immigration detainees, including asylum seekers, in prison.
>
> (ibid.: 68)

Borders of exception

We may want to turn, at this point, to Agamben's (1998) reading of sovereignty as that which operates in relation to the logic of *exception* where the distinction between inside and outside, inclusion and exclusion, bare life and political existence is collapsed. The space that emerges from within this zone of indistinction is (arguably) where the (bare) life of the asylum seeker/irregular migrant is caught up insofar as she or he is 'socially a "zombie"' (Diken, 2004: 87) whose existence is neither this nor that, whose presence is neither here nor there, as we discussed earlier. And what binds bare life to sovereign power, according to Agamben (1998: 109), is the notion of the 'ban', a relation of exception to the non-relational that is both inclusive and exclusive of life: 'He, who has been banned is not, in fact, simply set outside the law and made indifferent to it, but rather *abandoned* by it, that is, exposed and threatened on the threshold in which life and law, outside and inside, become indistinguishable' (ibid.: 28). Put differently:

> The ban attempts to show how the role of routines and acceptance of everyday life protects some over others, or how the protection of these others against themselves as the profound structure which explains the 'moment' of the declaration of exception.
>
> (Bigo, 2006a: 47)

So, from this perspective, when the Home Office *decides* to extend the power of detainee escorts to protect them and protect detainees (against) themselves, when it decides to expand the number of detention centres across the (edges of) country, when it decides to place asylum seekers under prison-like conditions, it is declaring a form of sovereign exception, or at least, *performatively* resuscitating 'a spectral sovereignty within the field of governmentality' (Butler, 2004: 61). The ban as such emerges as a by-product of such decisions and makes itself visible through correlative practices and spaces. It reveals the peculiar situation whereby detention centres exist both inside and outside the state: inside, for instance (but not only), in terms of their geographical presence that marks their 'inclusion' within the state's boundaries, and outside insofar as they are 'excluded' from its regular juridico-political procedures and structure. Detention centres are, in fact, the 'mezzanine spaces of sovereignty' (Nyers, 2003: 1080). Their exclusion from the official nation-state is an 'inclusive exclusion', and so is the case with the 'banned' and excluded life that dwells in these zones of exception. For it is the life of those who are '[c]onstituted as threshold political beings [...] defined precisely through their liminal status that places them on the outskirts of the community' (Zylinska, 2005: 93).

Hence, instead of assuming the total exclusion and exemption of asylum seekers and irregular migrants from the official political order, an Agambenian reading of the ban incites us to contemplate how these figures are subject to the rule of the law *while* being excluded from it and *aban*doned by it, how 'the sovereign exception does not leave in peace what it "abandons"' (Diken and Laustsen, 2005: 58), how through the instantiation of the state of exception in the form of detention centres, for instance, the sovereign law suspends itself while maintaining the norm in relation to the exception. As Rajaram and Grundy-Warr (2004: 34) put it: '[the] exemption must not be seen as a casting out of politics itself: the condition of exemption by which the normal political space is declared entails that the norm is dependent on its exempted other'. Or again: '[i]t is through the appropriation and control of the excluded, in effect its inclusion with its threat ameliorated, that the sovereign law maintains itself' (ibid.: 36) (in terms of *Secure Borders, Safe Haven*, this becomes apparent in the fact that the notion of legality/legitimacy vis-à-vis immigrants and asylum seekers is always invoked and articulated in relation to its opposite, and the image of the 'outsider' is always set as the backdrop for '(re)founding' the myth of 'belonging' and 'community'). In this sense, the refugee and the irregular migrant are regarded as part of the *system* of the nation-state whereby processes of control are set in motion in order to demarcate the lines between interiorised humanity and a remainder humanity (ibid.: 35). They represent the 'nomadic excess' (Diken, 2004: 85), which needs to be captured, managed and regulated. They are both a threat to the norm as well as an integral part of its perpetuation. Small wonder, then, the issue of asylum and immigration remains 'an acid test for politics' (ibid.: 83).

For Agamben, the ultimate figure of the ban is the *homo sacer* whose life is irreparably exposed 'in the relation of abandonment' (Agamben, 1998: 83). It is the life that is neither totally included nor completely excluded, neither protected by the law nor removed from it. As such, one might think (as Agamben and others do) of those clinging to speeding trains, climbing over border fences, crossing in leaky boats, walking in hellish deserts, smuggled in long-distance lorries (in order to reach the 'West') as *polymorphous* reincarnations of the homo sacer. But let us recall here the definition of the homo sacer. The homo sacer is 'the one whom the people have judged *on account of a crime*' (Agamben, 1998: 71, my italics) and is characterised by '*the unpunishability of his killing and the ban on his sacrifice*' (ibid.: 73, original italics). The first part of the definition is crucial insofar as it reveals the decisive difference between the figure of homo sacer and that of the asylum seeker/irregular migrant, for the homo sacer is bound (without redemption) to the 'diagram of crime' precisely for *committing* an act of crime. Whereas in the case of asylum seekers and irregular migrants, the diagram of crime is overridden (or at least supplemented) by a 'diagram of undesirability'[21] which, nevertheless, inevitably leads to the criminalisation of these figures (as in Eurodac and HAVANK), but a criminalisation whose content is devoid of the *act* of crime itself and filled instead with a pre-emptive supposition of '*unsavoury agency*' or '*dangerous agency*' (Nyers, 2003: 1070). As Bigo (2005a: 9) argues, the *dispositif* of the new carceral spaces such as detention centres generates 'the same carceral conditions as prisons, but without the legal judgment of guilt'. In this sense, while the figure of the homo sacer might be reproached for being too much of an exaggeration, it might equally be reproached for (at risk of adding to the irritation of the sceptics) underplaying the exceptional logic that underpins the treatment of asylum seekers and irregular migrants.

The second definition of the homo sacer (the characteristic of the unpunishability of his killing) is also important. We have mentioned before that the toughening of immigration and asylum policies (through technologies of border security, etc.) is exposing some people to extremely harsh experiences and increasing the possibility of their death while attempting to cross the borders of the West. By way of drawing a parallel, we may argue (and this remains open to contestation) that the death of asylum seekers and irregular migrants is nothing other than a *death through policy* which, like the 'unsanctionable' killing of the homo sacer, is treated neither as a punishable homicide nor as a sacrifice.[22] It remains true though that the Agambenian approach towards the figures of detention centres and homo sacer is hinging mainly on the *extreme* manifestations of sovereign exception and keeps feeding upon the pretence of generalisability. It thereby does not always capture the full picture nor does it account for the manifold ways by which the logic of exception is actualised, materialised and sustained. But what might prove helpful to us, at this moment, is to retrieve and retain the notion of 'inclusive exclusion' from these figures and examine its *enactment* through a less extreme example – an example that may also serve as a small corrective to the generalising and homogenising character of Agamben's dystopian account on the notion of exception, and add to the much needed empirical nuances.

Application Registration Cards

In October 2001, and part of the overhaul in asylum and immigration policy, the Home Office announced the introduction of ARCs for individuals claiming asylum in the UK (Telegraph, 30 October 2002). The ARC, also known as 'asylum smart ID card', is a biometric identity card containing the personal details of the asylum applicant (name, date of birth, photograph, etc.) and a memory chip with his/her fingerprints. Prior to the introduction of ARCs, applicants for asylum were issued with a Standard Acknowledgement Letter (SAL), a paper-based document of identification. Given its format, the SAL has been easily susceptible to forgery and counterfeiting (Home Office, 2002a: 54). As such, the aim of replacing SALs with ARCs is to provide a more reliable and tamper-proof means of identification by fixing the identity of the asylum seeker to his/her body in order to prevent the occurrences of double-dipping[23] and the 'abuse' of the welfare system. The scheme of ARCs coincided with the opening of several Accommodation Centres[24] where some asylum seekers can be provided with shelter, food, legal advice and other support services while their applications are being considered by the Home Office. Unlike the case in detention centres, residents of accommodation centres are free to come and go, but required to spend the night in the centres and report to police stations or other authorities on a regular basis. They also receive a small weekly cash allowance (£14)[25] and have access to health care and other benefits (Home Office, 2002a).[26] However, those who refuse to comply (by either leaving the accommodation centres or not carrying the ARC) are automatically disqualified from state support. The ARC must be presented in a number of situations and everyday transactions, including the reporting procedure, claiming support at the Post Office, accessing health care and so on. It is thus part of the overall strategies of tightening control over the provision of social services and benefits to asylum seekers, and ensuring that they do not 'disappear from the system'. In this respect, and to appropriate van der Ploeg's (1999a: 296) argument, 'the bodies of cardholders will become inscribed with their identities as [asylum seekers... and] implicated in the distribution of benefits, services, and rights'.

Here we have a pertinent example of *how*, in the management of asylum issues, the logic of inclusive exclusion can also function beyond the power paradigm of confinement, and in ways that cannot all too readily or uniformly be conflated with pure violence exercised outside of the realm of law – as in Agamben's formula. It is a case of managing the 'circuits of exclusion' (Rose, 1999: 240) by downloading mechanisms of *control* (Bigo, 2005a) onto a host of sites and into a cluster of technologies. Rose (1999: 240) argues that within contemporary forms of control, there are certain strategies that 'seek to incorporate the excluded [...] and to re-attach them to the circuits of civility' and others which 'accept the inexorability of exclusion of certain individuals [...] and seek to manage this population'. ARCs can be seen to be executing precisely, and concurrently, these very functions. For not only do they constitute, and indeed institute, the condition for gaining access to social services *as* an asylum seeker (inclusion within the nexus of sociality), but also demarcate the latter as an alien, a non-citizen, multiplying 'the possible loci of [inclusive] exclusion' (Rose, 1999: 243). The function of ARCs as a '(re)-attaching agent' is at once a function of 'attachment' as well as 'detachment', a function of 'inclusion' as well as 'exclusion': through his/her ARC, the asylum seeker is connected (precariously that is) to the order of civility only to be reminded that she/he does not belong to it, she/he is allowed to *perform* a certain form of inclusion only to *endure* another sense of exclusion. This double function of asylum smart ID cards is hence reminiscent of the fact that, in the field of asylum management, control strategies are not merely a matter of confinement, as is the case with detention centres, but also a matter of 'knowledge production' in the Foucauldian sense, which 'allows selection of thresholds that define [...] forms of inclusion and exclusion' (Ericson and Haggerty in Rose, 1999: 263).

A similar point can also be made with regard to accommodation centres. As mentioned earlier, the opening of these spaces is envisioned by the government as a solution to the problems (supposedly) caused by the 'dispersal system'. More specifically, they are considered as a response to the lack of social cohesion, believed to be brought about through the placement of asylum seekers within communities that are not very used to 'foreigners' and 'immigrants' (Telegraph, 29 October 2001). In this respect, accommodation centres can be regarded as an 'urban condom' (Sennett, in Diken, 2004: 92), designed to *prevent* the 'total' infiltration of asylum seekers into the interiority of the 'belonging' communities, in order to *pre-empt* the possibility of 'undesirable' intercontamination or subsequent need for forced 'abortionary' measures – a function that is very much in tune with the logic of control. Accommodation centres also trace the boundaries between those who are regarded as 'genuinely' potential refugees and those who are perceived as 'bogus'/'irregular' (economic) migrants (they are placed in detention centres instead). What becomes apparent, then, is that there are different (but co-constitutive) layers of inclusive exclusion, which are irreducible to any single monolithic conceptualisation (such as that of the figure of the camp). We should bear in mind, however, that the logic of accommodation centres does not completely nullify the logic of detention centres, but only smoothens it and opens it up to more complex and multidimensional mechanisms actualised through asylum smart ID cards, not least because the temporal freedom as well as the various benefits allotted to those residing in accommodation centres remain at the mercy of the whims of the state and always subject to their cancellation in case the terms and conditions (dictated by the government) are not abided to. For, as Diken and Laustsen (2005: 68) argue in Deleuzian terms, '[c]ontrol is a line of flight that escapes disciplinary entrenchment, but it has its own discontents, bringing with it nomadic forms of repression, and turning the freedom of movement into a new form of sedentariness'. In the case of asylum control management, this amounts to 'no ID card, no freedom' – perhaps not so much a line of flight per se, but a line of *descent* into the abyss of what Bauman (2002: 114) terms 'frozen transience'.

Having discussed the 'exceptional' deployment of biometric technology through the cases of Eurodac, border security and asylum smart ID cards, we can now circle back to the question of function creep and move on to examine how the use of this technology is increasingly being extended to the rest of the population.

When biometric identity systems are somewhat the rule

What is being witnessed through the proposals and implementation of *national* biometric identity cards systems, in various countries around the world, is a peculiar inversion by which practices that were previously confined to particular spaces and reserved to particular bodies are now overflowing and extending to the rest of the population. In other words, what was once an 'exceptional' treatment for those who are regarded as 'non-citizens', or those who end up becoming 'non-people', is now reaching those who are deemed to be so-called belonging, right-bearing citizens, and paradoxically, by the very same political order that determines their citizenship and with it their belonging. So, in a way, one can argue that biometric identity cards are the apparatus *par excellence* for controlling the population *en masse* while, nevertheless, maintaining and orchestrating a myriad of divisions within its overall biopolitical body.

However, this expansionary shift from the exception to the rule in the context of biometric ID cards systems is not to be regarded in mere *quantitative* terms. Nor should it be simplistically and reductively translated into the proposition that exception has become somehow generalisable in a single homogenous way. As this will inevitably result in overlooking the crucial and polysemic variations, nuances, paradoxes and ambiguities embedded within the rationale of biometric identity systems. What is needed instead is a *qualitative* stance able to confront these complexities and overcome both the dividing slash of 'rule/exception'[27] as well as the implied equation mark in 'exception *is* the rule'.[28] To do so, I will subscribe to Bigo's (2006a: 50) suggestion that exception is best understood as 'a specific form of governmentality' rather than as its contrary.[29] Taking up this suggestion will allow us to go beyond the problematic division between the 'Foucauldians' (governmentality) and the 'Agambenians' (sovereignty), a division whose first casualty is often the complexity and heterogeneity of the 'everyday itself' and its myriad technologies.

To suggest that exception is a form of governmentality is to, at once, undermine the extreme argument that exception is purely the aberrant suspension of the law and to challenge the oversimplifying belief that (neo)liberalism is the negating opposite of exception.[30] In other words, it is to convey that '[e]xception works hand in hand with liberalism' (Bigo, 2005a:18) and to take into consideration 'other visions of exceptionalism[31] that combine exception both with liberalism and the *dispositif* of technologies of control and surveillance which is routinised' (ibid.: 17). Foucault and Agamben are both needed here. And to invoke the two simultaneously, Bigo proposes the concept of the 'ban-opticon' which, as the term suggests, combines Agamben's use of Jean-Luc Nancy's notion of the 'ban' with Foucault's use of Bentham's model of the 'panopticon'.

What is interesting about this combination is that it is not merely a tautological 'conjunction of opposites', but stands as a useful heuristic device for understanding 'how a network of heterogeneous and transversal practices functions' (Bigo, 2006b: 35). In being so, it allows us to 'analyse the collection of heterogeneous bodies of discourses [...], of institutions [...], of architectural structures [...], of laws [...], and of administrative measures' (ibid.). In terms of our analysis of biometric ID cards, this collection is, beyond any doubt, of very high relevance. But before I start experimenting with the concept of the ban-opticon in relation to the spillover of biometrics and ID cards, allow me first to visit Foucault's notion of the pan-opticon and then briefly return to Agamben's reconfiguration of the notion of the ban.

Panopticism

In a chapter called *Panopticism*, Foucault (1975) outlines two major forms through which discipline and surveillance were exerted. The first being the *spatialisation* of the plague-stricken town by means of segmenting and immobilising space, and placing individuals within enclosures and under constant supervision. This mode of surveillance is based on a system of 'permanent registration' (registering the details of each inhabitant of the town) as well as on 'mechanisms of distribution' (in which each inhabitant is related to his name, his place, his body and his condition). The aim is to meet disease with order and eradicate any confusion that may emerge out of the 'mixing' of bodies, be they living or dead:

> The plague-stricken town, traversed throughout with hierarchy, surveillance, observation, writing; the town immobilized by the functioning of an extensive power that bears in a distinct way over all individual bodies – this is the utopia of the perfectly governed city.
>
> (Foucault, 1975: 198)

The second organisational form relates to the treatment of the leper, which, unlike 'the plague and its segmentations', functions by means of *separating* and *excluding* the leper from the healthy community through 'binary branding' (normal/abnormal, mad/sane, etc.) and 'exile-confinement'. From these two different images (plague and leprosy) that underlies the two different projects (segmentation and exclusion), Foucault goes on to explain the two ways of exerting (political) power: 'discipline' on the one hand (as is the case with the plague), and 'exclusion' on the other (as is the case with leprosy) (ibid.: 199). Despite the differences of these two modalities of power, Foucault insists that they are 'not incompatible ones' (ibid.). For power functions by way of *excluding* the 'infected' (here, the image of the leper stands as an emblematic figure of 'beggars', 'vagabonds', 'madmen' and so on, just as the image of the plague symbolises 'all forms of confusion and disorder') and *individualising* the 'excluded' so much so that lepers (all those who are symbolised by this image) are treated as plague victims (all those who are caught up within disorderly spaces). Power, as such, can be conceived as an amalgamation of both forms, and according to Foucault, Bentham's panopticon is 'the architectural figure of this composition' (1975: 200) *par excellence*.

Bentham's utilitarian plan for a prison, which is based on the principle of an observing supervisor placed in a central tower where he can see without being seen, serves as a compelling paradigm for the kind of surveillance that underpins the compound power of exclusion and individualisation. As Elden (2002: 244) explains, the model of the panopticon is precisely where the space of exclusion (of the figurative leper) 'is rigidly regimented and controlled' (as is the case with the figurative plague victim). The idea that 'visibility is a trap' (Foucault, 1975: 200) (i.e. the presence of the tall tower at the centre does not necessarily mean a supervisor is watching), and that power is 'unverifiable' (uncertainty about whether/when one is being watched), is what makes the model of the panopticon such a subtle and effective architectural apparatus, capable of automating and distributing the functioning of power. Power, in this sense, needs not to be enforced but merely 'internalised' through mechanisms of self-regulation. Such mechanisms render the observed as simultaneously the bearer (subject) of and the one subjected to power.

The model of the panopticon, however, is not merely a method of observation devoid of other disciplinary modes of power. Nor is it solely confined to the institution of the prison. Rather, it is also a machine that could be used to 'carry out experiments, to alter behaviour, to train or correct individuals' (ibid.: 203). It can be extended to a variety of other institutional spaces – ranging from schools and hospitals to factories, barracks and so on. It is, hence, the way in which this metaphor of the panopticon encapsulates different technologies and spaces of surveillance and discipline that Foucault places the notion of 'disciplinary society' under the umbrella of *panopticism* in an attempt to capture the diagrammatic strategies underlying power relations and in which 'positions' and 'identities' are some of the fundamental features of the functioning of 'panoptical' surveillance.

Importantly, panopticism, as Foucault understands it, marks the redundancy of the 'sovereign's surplus power' insofar as it is based on 'a machinery that assures dissymmetry, disequilibrium, difference [...] it does not matter who exercises power. Any individual, taken almost at random, can operate the machine [...] it does not matter what motive animates him' (ibid.: 202). *Framing* panopticism as such allows Foucault to maintain a clear-cut distinction between 'discipline' and 'sovereignty' and expel 'death' (that is so inextricably linked to sovereignty) from the realm of discipline (which he sees as being more about normalising and regulating the conduct of the living).

Nevertheless, as Diken and Laustsen (2005: 59) point out, 'it is possible to understand discipline as a new technique of abandonment, that is, as a technique through which sovereignty is still present'. For as we have seen with the *figure* of the leper, what is set in motion is precisely the power to *aban*don by way of excluding and confining. And according to Diken and Laustsen, '[a]s the prisoners internalized the gaze of authority, the citizens would internalize the risk of imprisonment' which means that [t]he ban could strike all; again, sovereignty reigns in *potentia*, omnipresent yet not necessarily real or actual' (ibid.: 60).

Diken's and Laustsen's remarks are certainly important in emphasising the non-obsolescence of sovereignty in discipline. But they also run the risk of being (mis)interpreted along the lines of what Rose (1999: 242) previously referred to as 'a dystopian and sociologized reading of the diagram of the Panopticon' in which discipline is regarded as 'a means of producing terrorized slaves without privacy' (ibid.). Interpreting these remarks in this way would not do much justice to them. For while the authors seem to maintain that the fear of imprisonment (or confinement in all its forms) is obdurately lurking behind the diagram of discipline, they also acknowledge that '[t]he primary goal of the panopticon was [...] not confinement but the manipulation of self-consciousness'; that '[t]he guiding idea is self-governance' (Diken and Laustsen, 2004: 61). Yet their conclusion does not follow Foucault's postulation in suggesting that discipline, through self-governance, renders sovereignty as obsolete, but puts forward, instead, the idea that sovereignty becomes omnipresent through 'invisibility', through the 'paradox of absence and presence', through 'the potentiality of abandonment' (ibid.).

What the authors overlook, however (and this is perhaps due to their over-reliance on a purely Agambenian reading of the ban), is that abandonment is not merely a matter of sovereignty and sovereignty alone, as Agamben wants us to believe. Rather, it remains, as Jean-Luc Nancy initially proposed, subject to various dynamics, interactions, and relations whereby the *mise-en-abîme* takes place not only under the sole command of the sovereign, but also according to how the 'community' react and relate to that which is to be excluded, and how forms of 'resistance' are formulated and mobilised. This argument is picked up most clearly in Bigo's (2005b) critique of Agamben's notion of the state of exception where he regards Nancy's concept of the ban as being more Foucauldian given its interest in the 'micro dimension' of the various dispositifs and disciplining assemblages.[32]

Banopticism

Let us now circle back to the dispositif of the banopticon. At first sight, it would seem that the rendering of biometric ID cards as a 'rule', applicable to the entire population is somewhat akin to a *hyperbolisation* of Bentham's panopticon. That is to say, a totalisation of surveillance whereby 'everyone' is placed under the cold gaze of the watcher, be it the state or otherwise. In fact, this perspective seems to dominate much of the discussions about the implications of biometric ID cards whereby the latter is often portrayed as a symptom of the rise of 'the Big Brother Society' (Hazlewood, 2006; Independent, 2 November 2006; Statewatch, 2005). It also has much resonance with the concerns surrounding the function creep of biometrics which, as mentioned before, tend to focus mainly upon the 'quantitative' spread of biometric surveillance. But as

Bigo (2006b: 35) argues, 'the surveillance of everyone is not on the current agenda'. The Pan remains a utopia: 'only the dream of a few agents of power, even if the rhetoric after September 11 articulates a "total" information' (ibid.), and even if biometrics is, to some extent, a 'realization of Bentham's dream [;] of society where each citizen is tattooed with their name' (Walters, 2005: 9). What seems to be mostly at issue is, instead, the notion of the Ban; that is, the surveillance and control of 'certain' groups and the *normalisation* of the majority (Bigo, 2006b: 35), which makes 'surveillance more refined and precise, rather than extending its general reach' (Bigo, 2005a: 58). This way, the ban keeps the surveillance machine 'from getting jammed' (ibid.: 39) by inducing 'a kind of prudent relation to the self as condition for liberty [normalisation]' while refining 'the criteria for inclusion and [specifying] them at a finer level [selectiveness]' (Rose, 1999: 243).

And, in terms of biometric identity systems, this selective banoptical capacity, or the 'banoptic sort' to redefine and update Gandy's (1993) term, can be achieved through the various technologies which form the identity system (biometrics, information and communication technology, identity registers, smart cards, etc.). For their functional specificities allow for the manipulation of the system's interactivity and exploiting the features of database relationality and mining. All these techniques facilitate the assembling of public and private files and the linking between disparate systems and databases. This banoptic sort can also be achieved through the *administrative convergence* and compulsory *cross notification* which allow (and impose) the merging and sharing of personal information across a plethora of public and private bodies.

Taken together, these techno-administrative processes provide a sophisticated platform for 'abstracting' information from circulating bodies and fitting it into 'neat categories and definitions' (Adey, 2004: 502), while enabling the possible 'fragmentation' and 'distribution' of data across a multitude of searchable database records. To the extent that '[p]rofiling is the ability for information or data about an individual to be built up' (ibid.: 505), these processes have, therefore, the technical potential to create various *profiling* mechanisms and deductive classifications in order to systematically 'sort among the elements to formulate what and who must be surveyed' (Bigo, 2006b: 39). As such, national identity cards systems are emblematic of the increasing shift towards 'pro-active governmentality' (ibid.), towards 'the pro-active surveillance of what effectively become suspect populations' (Levi and Wall, 2004: 200). It is one way of enacting contemporary control practices at a larger social scale, but also 'selectively', through processes that are 'dispersed' rather than unified, 'rhizomatic' rather than hierarchical, 'relational' rather than totalised (Rose, 1999:246) – without, however, disposing of the all-too-familiar disciplinary mechanisms.

To this end, just like the deployment of biometrics at the border can function as a mechanism of triage whereby some identities are given the privilege of smooth passage whereas other (non)identities are arrested (literally), so too can biometric ID cards systems in that they inevitably 'reinforce the advantages of some and the disadvantages of others' (Bigo, 2006a: 57). So, perceiving this technology of identification along the lines of the banopticon allows us to understand its function as an *inner, portable, mobile* and *(im)mobilising* border that is capable of creating caesuras within the totality of the 'population-body' and subsequently paving the way for biopolitical sorting and (sub)divisions. This approach also allows us to reveal that the intensification in the scope of control through biometric ID cards does not have the 'same' impact on everyone, but remains very much a matter of 'conditional access to circuits of consumption and civility, constant scrutiny of the right of individuals to access certain kinds of flows of consumption goods: recurrent switch points to be passed in order to access the benefits of liberty' (Rose, 1999: 243). As such, the banoptical trait of this technology is inscribed into the overall logic of liberalism which seeks to regulate and govern the flows and transactions of the population by creating 'self-managing citizens' (ibid.) on the one hand, and 'controlling excessive freedom' (Huysmans, 2006: 97) on the other.

And again, just like asylum seekers' ARCs are designed to control access to social benefits and other services, so too are biometric identity cards. This spillover from the 'asylum body' to the 'national body' via biometric cards systems is more than a matter of extending the reach of available technology to meet surveillance needs, and more than a symbolic and material expression of (territorial) sovereignty. It is an attempt to sustain the *immanent* and *fluid* functioning of the liberal economic machine by opening up the population-body to powerful institutional and bio-technological processes which 'govern everyday practices on the basis of routines, diagrams [...] regulations and devices that categorize and control excessive conduct of freedom' (Huysmans, 2006: 97). Such banoptical processes are actualised in a 'double assemblage' (Bogard, 2006: 105) which combines what Deleuze and Guattari refer to as 'machinic' and 'enunciative' assemblages (in ibid.). The machinic assemblage of biometric ID cards is the 'content' which subsumes those technical elements and specificities we discussed above, namely the interconnected databases, identity registers, the 'card' itself, and all other material components that facilitate the distributions of bodies and flows across networks. Whereas the enunciative assemblage is the 'expression' by which (truth) statements and claims are produced in order to attach meanings and values to bodies and techniques. For this reason, it is not enough, as Bogard (2006: 107) reminds us, to only examine what the assemblage *does* (as seen earlier, the 'does' element of technology is precisely what permeates most of the debates on the spillover of biometrics, leaving other elements untouched). Instead, one should also look at what the assemblage *says* – or rather what it is *made* to say.

The 'said' of the assemblage, in the context of biometric ID cards systems, is articulated at a 'discursive' level by means of linguistically placing '*the* nation' in 'an existentially hostile environment and asserting an obligation to free it from threat' (Huysmans, 2006: 50). For instance, the *problematisation* of the issue of globalisation and the scale of movement in terms of inevitable threats from various figures recasts socio-political problems as 'the problem of the [dangerous] excluded' (Rose, 1999: 258) while framing solutions in terms of available technologies rather than the real underpinning problems (Bigo, 2006a: 56). In this way, governments derive the legitimacy to introduce biometric ID cards from the production and distribution of epistemological otherness and dangerousness, and through 'the simulacrum of a politics of proximity designed to reassure the good citizens, and the zero tolerance designed to deter the rest' (Bigo, 2006b: 39). These enunciative techniques infuse a sense of prudentialism which urges 'citizens' to perform 'responsible self-government' (Rose, 1999: 259) by embracing the 'benefits' of identity systems (more on this in Chapter 4). They give identity cards 'a vampyric "parasitic vitality" through attachment to something valued by the public' (Agar, 2001: 119) (e.g. the welfare system, security). As such, biometric ID cards become the ontological and epistemological foundation for exercising (and fixing) identity, citizenship, and belonging. They become a means of normalising the majority while weeding out certain groups on the basis of their perceived threat and potential behaviour. In this sense, the spillover of biometric ID cards should not only be considered in light of its 'repressive' effects on those who are excluded from the liberal sphere of free movement and accessibility, nor merely in light of the welfare chauvinism which seems to dominate much of the contemporary policies of the (over)developed countries. But it needs to be considered also in light of its 'normative and productive dimensions' that run through the entire population-body and in which '[w]ords and things are [...] held together by [...] relations of power' (Bogard, 2006: 107). And, power, as Bigo (2006b: 42) argues in Foucauldian terms, 'is not only repressive. It induces and produces modes of behaviour'.

At the same time, *within* this liberal space of governance, and somewhere between the imperative of self-management and the control of excessive freedom, lies in the notion of exceptionalism: the other feature of the banoptical dispositif. This vision incites us to keep in mind that exception (whether sovereign or *otherwise*) is not all too absent from contemporary diagrams of control – no matter how liberalism seeks to legitimise itself through 'the idea of the separation of powers by which power is supposed to limit itself, particularly through checks and balances, with the effect that the population finally actively consents to be an accomplice of its own domination and to rely on "justice" and lawyers for its "freedom" ' (ibid.: 36–7). It reminds us that the excluded upon which the intensity of surveillance is inflicted is precisely what determines the limit that demarcates the 'porous' contours of the inside. In other words, the *figure* of the fraudster, the terrorist, the criminal, the double-dipper, the illegal immigrant, the asylum seeker and so on, is precisely what allows the establishment of the value-laden figure of the citizen itself with all its accompanying conceptions of interiority, legitimacy, rights, responsibilities and so on. The 'dangerous outsider' and the 'belonging insider' are but co-constitutive entities.

As a guiding thought for understanding the close relationship between exceptionalism and biometric identity cards, it is worth imagining 'the circular relation that characterizes political constitutive acts: security claims introduce the existence of larger political units – i.e. "nations" or "larger groups" – by identifying them as being under threat while simultaneously asserting that the unity is born out of the presence of the threat' (Huysmans, 2006: 50). In this circular relation, 'low politics' are dramatised, overemphasised and rendered as a matter of urgency and priority so as to be elevated to the status of 'high politics' (van Munster, 2005a: 3). This is done in a way that notions such as 'public security and order', 'state of danger' and 'case of necessity', 'which refer not to a rule but to a situation, [render] obsolete the illusion of a law that would a priori be able to regulate all cases and all situations' (Agamben, 1998: 172). High politics, as it were, derives its significance precisely from its ability to decide on the exception by moving 'certainty' and 'calculability' outside the juridical rule and making it possible to stimulate and instantiate immediate and exceptional (but nonetheless permanent) actions vis-à-vis a particular threat – be it *actual*, *potential* or even *imaginary*. And what inheres in the passage from low politics to high politics is a problematic intersection between inside/outside, self/other, identity/otherness, us/them, safety/danger and so forth to the point of indistinction, inasmuch this passage itself is but a limit-concept by which such divisions pass through one another and the move to exceptional measures is made possible. Contra Agamben, however, and at least in the context of biometric ID cards systems, the law needs not be *suspended* in order to give way to exceptional measures. Instead, the law undergoes specific derogations admitted by the law itself under certain circumstances (Bigo, 2006b: 54). Or as Huysmans and Buonfino (2006: 10) explain:

> The policies that are discussed may be exceptional but the politics through which these policies are supported and contested are not. The way the debate is conducted, the mobilisation of interests, etc. does not exceed the boundaries of the 'normal' institutionalised way of doing politics.

So although the proposals and implementation of biometric ID cards systems are partaking of exceptional security measures, they remain, also, subject to the usual chain of political manoeuvring in a liberal democracy. This can be seen though the forming of various committees and the commissioning of a number of studies (e.g. the *LSE Identity Project*, the House of Commons Science and Technology Committee's *Identity Card Technologies: Scientific Advice, Risk and Evidence*, the European Commission's *Biometrics at the Frontiers: Assessing the Impact on Society*) with the view to provide 'expert' advice to the government on the implications of this technology. Added to that is the inclusion of public opinion polls as well as the discourses of human rights groups, and so on into the overall process of debating the introduction of biometric ID cards systems. The existence of these thought communities, is precisely what preserves a sense of 'normality' at the heart of liberal policies. At the same time, however, this existence by no means amounts to the cancellation of the exceptionalist impulse in policing and governing. For it signals towards the 'normalization of emergency as a technique of government by unease' (Bigo, 2006a: 63), a process that intrinsically carries within it potential illiberal practices (such as those of profiling and criminalising particular targeted groups).

Conclusion

Finally, framing the problem of function creep of biometric technology and identity systems in terms of the notion of exception allows us to go beyond the reductive and technologically deterministic approach in which the debate is merely reduced to what technology itself does and does not. It allowed us to open up discursive and multidimensional ways of comprehending the potential impact of this technology on embodied existences while stressing upon the heterogeneity and polysemic nature of the scheme. And, it allows us to draw a more general conclusion wherein the inversion in the role of biometric identity cards systems can be understood as a peculiar and paradoxical move by which, on the one hand and at the ontological level, the categories of the citizen and the non-citizen, the legal and the illegal, the belonging and the alien and so forth are rendered indistinguishable (this, in terms of the scanning of bodies in their *totality*[33]). Whereas, on the other hand, and at the socio-political level, identity cards may be deployed as the means by which these very categories can be established, *imagined* and differentiated (so as to demarcate the boundaries between those who can be included and those who should be excluded, and decide on the administration and distribution of services, entitlements and rights). This paradoxical role of identity cards systems is one of the epitomes *par excellence* of the aporia at the heart of modern biopolitics in which, and as Foucault (although hesitantly), Arendt and most explicitly Agamben have articulated, the processes of totalitarianism and democracy are concurrently brought into play and life itself is made both the *subject* and the *object* of (bio)political calculations and decisions. To probe even deeper into the aporia of biometric identity systems, we need to take a closer look at the concepts of 'identity', 'security' and 'citizenship', which are the themes of the next chapters.

3
Recombinant Identities: Biometrics and Narrative Bioethics

Historically, and whether at the micro (individual) or macro (social, national, cultural, etc.) level, the notion of identity has often been bound up with that of conflict or crisis. Contemporary articulations and practices of identity are no exception. They are increasingly being marked by what Anthony Giddens (1991) refers to as 'ontological insecurity', that is, a deep sense of anxiety and uncertainty about the question of 'who someone is' in relation to oneself and to others, be they other 'individuals' or institutions. Rightly or wrongly, out of convenience or out of paranoia, identity is now routinely being problematised in terms of risk, or more specifically, as being *at* risk; the risk of fraud, the risk of crime, the risk of terrorism, the risk of illegal immigration, the risk of illegal working and so on. Within the current policy debates and discussions regarding biometric technology and identity systems, the age-old question of 'who is who?' continues to occupy a central stage, not only because of its highly political relevance, especially to issues relating to the much contested domain of membership and the attribution of rights and obligations, but also because of its inherent and irreducible ambiguity that poses a challenge to the ongoing technical attempts to find a definitive and fixed answer to it. As a response to such challenges, various techniques and technologies have been mobilised with the aim to protect and manage the uniqueness of identity. Among the most notable of these techniques is the securitisation of identity through biometric technology.

As discussed in the previous chapter, biometric technology has recently witnessed a massive growth and a rapid proliferation within many areas of society. Its application, which was traditionally reserved for particular practices, is now covering a broad array of spaces and functions, ranging from border control and asylum regulation to the management of social services and medical records. Unsurprisingly, this expansion in scale and deployment has raised many concerns over the potential ethical implications of biometric technology. The majority of these concerns, however, remain largely framed within the normative discourses of privacy, liberty and data protection, leaving aside other issues that are by no means less pertinent to the political and ethical analysis of the use of biometrics as a means of identification and identity verification.

In this chapter and remaining with the example of asylum, I shall address one specific aspect of the 'bioethics of biometrics', an aspect that – despite its fundamental relevance, and with a few exceptions (Aas, 2006; Ceyhan, 2008; Lyon, 2008; van der Ploeg, 1999b) – has not yet managed to secure the space it deserves within the academic literature on biometrics and its implications. This aspect relates mainly to the relatively basic and commonplace, but also highly problematic and notoriously intricate, question 'who are you?', which, in my view, constitutes an interesting backdrop against which one may start delineating the distinctive bioethical characteristics of biometrics beyond the familiar trope of privacy and the like. For it encapsulates the ontological and epistemological challenges of uniqueness and identity that biometric technology aspires to respond to and manage. Inevitably, addressing such a question in relation to biometrics requires us to enquire, first and foremost, into the ways in which biometrics is *about* the uniqueness of identity and into the *kind* of identity biometrics is concerned with. One way into this enquiry is to be found in the question of 'identity itself'.

The identity in question

> Everyone's unique. Let us keep it that way.
>
> (Home Office, 2008a)

In a sense, and at least at the systematic and structural level, recent attempts to securitise identity through biometric technology seem to have, as one of their main tasks, the 'simplification' of the meaning and function of identity. They are underpinned by scientific discourses and practices that tend to convert the subjective and, in many ways, profound dimensions of identity into hyper-empirical and objective programmatic Boolean operations of true/false and positive/negative. Their overarching aim is to purify, so to speak, the articulations of identity from ambivalence and instability while rendering them immune to the problems associated with 'human fallibility' (Gates, 2005: 38), which technically, and for so long, had made the process of identification by and through human agents/subjects a rather inefficient and unreliable enterprise. Doubtless, however, and despite such attempts, identity continues to be a highly contextual, elusive, malleable, ubiquitous and, indeed, 'complex' concept. Therefore, it does neither lend itself easily to definition nor remain unchangeable. As such, any discussion about identity and its securitisation needs to be at grips with some of the variations in the meaning of identity itself.

'Controversies about personal identity are as old as Western philosophy, not to cite Buddhism and Hinduism' (Mordini and Ottolini, 2007: 51), and defining who someone is has always been a major preoccupation of metaphysics. Nevertheless, the majority of philosophical discourses remain, as Arendt and others argue, 'unable to determine in words the individual *uniqueness* of a human being' (Kottman, 2000: vii) inasmuch as this uniqueness 'retains a curious intangibility that confounds all efforts toward unequivocal verbal expression' (Arendt in ibid.). In other words, who someone *is* escapes the confines of language and the boundaries of definitions, challenging any attempt to complete linguistic appropriation. For this reason, 'the moment we want to say *who* someone is, our very vocabulary leads us astray into saying *what* he is' (ibid.): for example, his qualities and attributes which qualify him as an individual, a citizen, a member – 'as if the task were simply to fill in the content of [...] personhood' (Butler, 2005: 31). Or, as Caplan and Torpey (2001: 3) suggest, in the context of identity documentation, 'the question "*who* is this person?" leaches constantly into the question "*what* kind of a person is this?"' (my italics). This collapse of the 'who' into the 'what' within the philosophical discourses of personhood and identity, as well as within the practices of identification, indicates their inherent limitations in capturing the ambiguity of identity and the complexity of the lived experience. It is also indicative of 'the extent to which traditional philosophy and politics respond to universals, rather than to unique persons and their interaction' (Kottman, 2000: ix).

As a response to these limitations, various efforts have been devoted to developing more nuanced and inclusive accounts that take into consideration the ambivalent and double-sided character of identity without conflating the 'what' and the 'who' aspects. In *Relating Narratives: Storytelling and Selfhood* (2000), Adriana Cavarero, for instance, provides an interesting take on the question of identity by foregrounding the importance of the notion of 'narration' which, according to her, enables the disclosure and preservation of the uniqueness of each life.

Inspired by the work of Hannah Arendt, Cavarero locates the 'what' element of identity within the realm of philosophy, and the 'who' element within the realm of biography. She perceives the relation between the two as that of

> [a] confrontation between two discursive registers that manifest opposite characteristics. One, that of philosophy, has the form of a definitory knowledge that regards the universality of Man. The other, that of narration, has the form of a biographical knowledge that regards the unrepeatable identity of someone.
>
> (Cavarero, 2000: 13)

As such, Cavarero differentiates between the biographical or 'narratable self', which is marked by and formed through the experience of storytelling, and the traditional 'subject' as known throughout the metaphysics of subjectivity, with its accompanying concepts of individuality, agency, control and so on. Whereas the latter is continuously caught up within the philosophical persistence of 'capturing the universal in the trap of definition', the former emerges out of the revelation of 'the finite and its fragile uniqueness' through the delicate art of narration (ibid.: 3). And, through narration, the self is constitutively and continuously *exposed* to others. This exposure, according to Cavarero, is precisely what reveals the singularity and 'whoness' of a person, and makes the social and political life possible. The uniqueness of personal identity, in this sense, is not that which can be derived from a *universal* substance (being *a* human, for example) nor reduced to the *particular* 'whatness' of the person (having this or that attribute or belonging to this or that category[1]). It is rather of a totally 'expositive', 'exhibitive' and 'relational' character so much so that '*who* each one is, is revealed to others when he or she acts in their presence in an interactive theater where each is, at the same time, *actor and spectator*' (ibid.: 20–2).[2] Hence, even the act of telling one's *own* story is very much dependent on the existence of *necessary others*. In advancing such an argument, Cavarero is challenging not only the supposed sealed 'interiority' of the subject that characterises the individualist doctrine, but also the autonomy of traditional *auto*biography whereby the self turns itself into an 'other' in order to tell his own story. This 'other', for Cavarero, is merely 'the fantasmatic product of a doubling, the supplement of an absence, the parody of a relation' (ibid.: 84). In contrast, Cavarero's other is 'really *an* other' whose existence and presence are necessary for recognising and designating the uniqueness of the self:

[I]n the uniqueness of the *who* there is no homage to the self-centered and titanic subject of romanticism. The *who* does not project or pity herself, and neither does she envelop herself within her interiority. The *who* is simply exposed; or, better, finds herself always already exposed to another, and consists in this reciprocal exposition.

(ibid.: 89)

Another useful place, where different concepts of identity are delineated, can be found in the work of Marya Schechtman (1990). Schechtman draws a distinction between the *question of reidentification*, as known in psychological-continuity theories and which involves the elucidation of 'the necessary and sufficient conditions for saying that a person at time t_1 is the same person as a person at time t_2', and the *question of self-knowledge*, which refers to the set of beliefs and experiences that are expressive of *who* the person is (ibid.: 71). So, while the first question is concerned with the notion of 'sameness' over time and space, the second question looks at the 'uniqueness' of the person. This distinction is demonstrated by Schechtman in the following way:

The question 'Who am I?' might be asked by an amnesia victim or by a confused adolescent, and requires a different answer in each of these contexts. In the former case, the questioner is asking which history her life is a continuation of [reidentification], and, in the latter, the questioner presumably knows her history but is asking which of the beliefs, values, and desires that she seems to have are truly her own, expressive of who she is [self-knowledge].

(ibid.)

Like Cavarero, but through a different vocabulary, Schechtman argues that contemporary (analytical) philosophical accounts on identity have been predominantly focused on the question of reidentification, disregarding the component of self-knowledge which, she believes, is an integral part of one's coherent self-conception and sense of personal identity. She also suggests that the dead-end encountered by psychological-continuity theorists vis-à-vis identity is largely due to the conflation of these two questions (ibid.: 72) (just as the conflation of the 'what' and the 'who' aspects of identity is what marks the irremediable limitations of philosophical discourses of identity). In this sense, Schechtman emphasises the importance of attending to the question of self-knowledge while addressing the issue of identity. However, and unlike Cavarero's narratable self, which attempts to break away from the metaphysics of subjectivity, Schechtman's articulation of identity as self-knowledge seems to be confined within this very metaphysics. As such, Schechtman's approach, as opposed to Cavarero's, pays little attention to the importance of the notions of exposure and otherness in contributing to the *process* of self-knowledge.

At this point, one might even argue that the clear-cut differentiation between reidentification and self-knowledge is not as pure and absolute as it may seem; and that trying to maintain a sharp line of demarcation between these concepts runs the risk of resuscitating some undesirable forms of dualism. For, in a concrete sense, such concepts constantly leak into each other, not least because of the ways in which the experiences of 'embodiment' and the practical 'performance' of identity in everyday life remain a matter of continuous 'contamination', given their socio-cultural and political embeddedness. van der Ploeg (1999b: 40) raises a similar argument while framing Schechtman's two concepts of identity in terms of the difference between a third-person perspective (entailed in the concept of reidentification) and a first-person perspective (involved in the question of self-knowledge). She asserts that the absolutisation of this difference is underlined by the unwarranted assumption that 'there is something like an authentic, true self to which the subject has an exclusive, epistemologically privileged access. This ignores the social and cultural dimension in identity formation of even the most "private" self' (ibid.). And it is precisely this assumption that Cavarero's approach attempts to overcome through the constitutive inclusion of the other in the process of narration – or put otherwise, through the intertwining and fusion of different person perspectives. Atkins explains a similar interrelation in the following way: 'who a person is is the named subject of a practical and conceptual complex of first, second and third-person perspectives which structure and unify a life grasped as it is lived' (Atkins, 2004: 347). Correlatively, even Cavarero's distinction between the who and the what aspects of a person is not to be regarded as a sharply dichotomous one: who someone 'is' is surely affected, to some degree, by what he or she is – even when this 'what' element remains indifferent to the bewildering whoness and uniqueness of the person. In other words, while the 'story' and the 'attributes', the 'who' and the 'what', are by no means the 'same', they do, however, interact *beyond* a binary or mutually exclusive relation.[3]

Nevertheless, and for the sake of analysis, maintaining a distinction (at least a relative and contingent one) between the question of 'reidentification' and the question of 'self-knowledge', between the question of 'who' and the question of 'what' in relation to the notion of identity, may help us turn the puzzling problematic of who someone is into an (ethical) opportunity for understanding what sort of identity biometrics is concerned with mostly, or as van der Ploeg (1999b: 39) puts it, 'in what sense "identity" is at stake in biometric identification techniques'.

Reconfiguring identity through biometric technology

Traditionally, and as far as the process of identification is concerned, there are three major sets of characteristics that are used to identify and describe a person:

- *What* she is (face, voice etc.)
- *What* she knows and uses to identify herself (name, address, social security number, etc.)
- *What* she has that provides for recognition of her identity (passport, token, etc.).

(Carblanc, 2009: 12, my emphasis)

There is a clear sense in which the 'remediation'[4] of these three vectors of identity through the introduction of biometric technology retains a fundamental interest in the 'what' element of a person, be it in terms of the use of physical attributes (what one *is*) or the convergence of indexical data (what one *knows*) and biocentric data[5] into biometric documents of identification (what one *has*). So in this respect, one might be tempted to argue that the relationship between biometrics and identity takes, or rather *maintains*, a narrow dimension vis-à-vis the question of 'who someone is', to the extent that it is based upon the reduction of the person to her 'whatness'. Similarly, it can also be argued that biometrics is primarily concerned with the question of 'reidentification' in which notions such as continuity, coherence and sameness are of utmost importance. Schechtman (1990: 71) explains that

> The primary contenders for a criterion of personal identity have been the bodily criterion and the psychological criterion, which are based, respectively, on the intuitions that it is sameness of body and sameness of personality which are responsible for sameness of person.

'Sameness of body', as it were, conceives the body itself as a 'constant', able to guarantee a certain degree of continuity and stability across time and space. This type of sameness is precisely what biometric technology is interested in – at least in the technical sense. Sameness of personality, on the other hand, involves, to a large extent, the precarious and difficult 'achievement' of a coherent personality that is itself very much reliant on the continuity and coherence of *subjective* experience. And, as Mordini and Ottolini (2007: 51) point out, '[t]he problem arises when we try to understand whether the subjective experience of this coherent personality corresponds to any real object or is just a useful figment'. In this regard, biometrics appears as a means of circumventing this 'problem' by finding recourse in the body itself and turning it into a stabiliser of identity, and by shifting the question of identity from the domain of narrative (the story of who someone is) to that of templates (digital samples of one's biological data).

Parenthetically, however, it is not that the body is absent from the second notion of sameness, that is, sameness of personality and its relation to subjective experience. Quite the contrary. The body, as we learnt through the different strands of phenomenology and the extensive feminist literature (and indeed through our own personal experiences), is an integral part of one's experience and awareness of being-in-the-world. But there remains a crucial difference in terms of the ways in which the body itself is perceived in both sameness of body and sameness of personality. At risk of oversimplifying, we can postulate that in the first model of sameness, the body has the status of an 'object' amenable to abstraction, measurement, digitisation, storage, distribution and so on. The relationship between identity and the body in this instance is of an 'external' order. That is to say, the person is regarded as 'having' a body that remains the same throughout life and 'upon' which many activities can be exercised (biometric identification for instance), whereas in the second model of sameness, the body is regarded as a subject 'through' which the world is lived and experience is made possible. Atkins (2000: 337) argues in phenomenological terms that 'there can be a lived *world* only because *my body* is itself part of the world which it experiences'.

The latter model has much resonance with what Paul Ricoeur (1992) refers to as *ipseity*. Ricoeur situates the notion of identity within the dialectic of *idem* and *ipse*; *sameness* and *selfhood*. Idem-identity involves something similar to that which is implied by the notion of sameness of body, particularly in its consideration of the body as a constant entity that can be compared to other entities outside time variants. It corresponds to 'the notion of identification, understood in the sense of reidentification of the same, which makes cognition recognition: the same thing twice, *n* times' (Ricoeur, 1992: 116).[6] In so doing, idem-identity assumes some principle of 'uninterrupted continuity and permanence in time' (ibid.: 117). It can take the form of *'numerical identity'* which indicates oneness and unity as opposed to plurality and diversity (e.g. passport or ID card number), or *'qualitative* identity' which stands for extreme resemblance and interchangeability (e.g. x and y wearing identical clothes) (ibid.: 116, 122).[7] For Ricoeur, this version of identity, which takes as its premise the sameness of body and the cardinal notion of reidentification, inevitably results in the increased concealment of selfhood. 'And this will be the case as long as the characteristics related to possessive pronouns and adjectives ("my," "mine") have not been connected to the explicit problematic of the self' (Ricoeur, 1992: 33). That is to say, as long as the relation of body to identity remains contained within and reduced to an external order of ownership, that is, 'having' a body.

Ipse-identity, on the other hand, is about selfhood and involves the biographic, embodied, temporal and narrative dimension of who someone is. Rather than being an emblem of constancy or a datum of sameness, the body, in ipse-identity, is regarded as an *attestation* to selfhood itself – as 'the most overwhelming testimony in favor of the irreducibility of selfhood to sameness' (ibid.: 128). Much like Cavarero, Ricoeur acknowledges the vital importance of otherness and the constitutive role of relationality to the formation and (narrative) formulation of ipseity. He also lodges similar complaints against 'cogito philosophies' and metaphysical discourses of identity in terms of their substitution of the question of 'who' for the question of 'what' and the ensuing eclipsing of the question of selfhood and its uniqueness. To this end, Ricoeur regards the self-attesting dimension of ipseity as a means of protecting the question of 'who' from such a misleading substitution. He writes, '[i]t is self-attestation that, at every level – linguistic, praxic, narrative, and prescriptive – will preserve the question "who?" from being replaced by the questions "what?" or "why?" Conversely, at the center of the aporia, only the persistence of the question "who?" – in a way laid bare for lack of response – will reveal itself to be the impregnable refuge of attestation' (ibid.: 23). In this sense, then, attestation, in all its polysemic and polymorphous forms including those of narrativity and embodiment, is very much reliant on whoness for its own actualisation and subsistence, just as the question of 'who' remains dependent on attestation for its own revelation and survival. This binding kinship between the two is precisely where the ethical plane unfolds, according to Ricoeur.

From all the above considerations emerges a series of intricate questions, questions that cannot be sidestepped if we are to understand the relationship between biometrics and identity – especially if we assume the phenomenological inseparability of body and identity: where does the 'biometric body' stand here? Does it merely belong to the realms of the 'what' and the 'idem'; or does it straddle both the 'who' *and* the 'what', the 'idem' *and* the 'ipse'? Is it merely an 'object' of abstraction, comparison, matching and reidentification; or does it gesture towards a less reductionist and a more complex vision?

The biometric body

To be sure, the (re)turn to the body for the establishment of identity in biometric technology seems almost like an ironic twist vis-à-vis Cartesian dualism. For while the Cartesian imaginary is underlined by the (erroneous) belief that consciousness is detached from the body, that the body has little relevance to identity and that it is an impediment to objectivity, biometric technology, on the other hand, lays claim to the idea that identity can 'objectively' be determined through the body and in ways that are somewhat independent of consciousness.

> En ce XXIe siècle, le corps prend sa revanche. C'est à lui que l'époque moderne confie la tâche de livrer l'identité de la personne, de dire qui est qui et qui, par conséquent, a le droit d'entrer.
>
> (Valo, 2006: 21)

> [In this 21st century, the body takes its revenge. It is in the body that the contemporary epoch entrusts the task of delivering personal identity, to say who is who and who, as a result, has the right to access.]
>
> (my translation)[8]

One may quibble here about whether this reversal of status is truly a 'revenge'. In a slight sense, it is, insofar as, 'biometrics gives the body unprecedented relevance over the mind' (Aas, 2006: 154). 'I think therefore I am' becomes 'I am I' (Lash in ibid.: 155) or rather, 'I am that' (*that* name, *that* fingerprint, *that* hand pattern, *that* face scan, etc.), where 'I' is heavily reliant on the body, and its algorithmic representation, to assert its (official) identity. And, instead of being relegated to the status of the 'container of the soul' as in Cartesian dualism, the body is now being treated as the forensic dust of identity, as the crystal ball through which the 'astrologists of identity' seek to predict potential risk and future dangerousness. The body, as such, is increasingly regarded as 'a source of instant "truth"' (Aas, 2006: 154) encapsulated in the expression 'the body does not lie', a catchphrase that has so conveniently been marketed by biometrics industry. But this instant truth is merely a truth *about* the body *qua* body-data. It is a truth that excludes the 'tale' of the body, that is to say, its narrative and biographical dimension, without which a person can hardly maintain a sense of whoness and (temporal) coherence.

In fact, the entire philosophy of biometric technology is based upon an epistemic suspicion towards the 'story'. It is based upon the belief that 'the mind is deceiving while the body is "truthful"' (ibid.). For this reason, when the biometric body speaks, it speaks in a language that silences the biographical story of the person whose body is ordered to speak. It therefore occludes the 'echo' of whoness while merely revealing the 'trace' of whatness. As Aas (2006: 154) explains,

> A talking individual, who owns the body, is in fact seen as unnecessary and, even more importantly, insufficient for identification. Now only the body can talk in the required ways, through the unambiguous and cryptic language of codes and algorithms. When a body provides the password, a world of information opens. Databases begin to talk. On the other hand, when the individual talks, the words are only met with suspicion.

So, in this respect, although biometrics seems to be reversing the internal order of Cartesian dualism by giving supremacy to the body over the mind, it is still sustaining, to some extent, a similar dualism between the two by doing just that. If Cartesian dualism, as we know it, has a tendency to disregard the fact that mind requires body, biometric dualism has a tendency to disregard the fact that body requires mind. According to Mordini and Ottolini (2007: 54), '[b]ody requires mind, not in the trivial sense that you need a neurological system to animate the body, but in the profound sense that the very structure of our body is communicational [...] We do not just need words. We are words made flesh'. In this regard, biometrics can be considered as yet another instance whereby the unity of mind and body is negated. And, although biometric technology recognises the fact that bodies are indeed biographies, it hardly offers an outlet for listening to those biographies. This is because the knowledge it produces is not based on 'mutual communication', but on 'one-way observation. It is clearly knowledge marked by a power relation' (Aas, 2006: 153).

Furthermore, this reversal of status does not necessarily amount to the body's escape from the status of the object. For although biometric technology places bodies centre stage, these bodies are 'already defined merely in terms of their sameness to other data [...and] other bodies' (Lyon, 2008: 507). As mentioned earlier, establishing sameness of body is a paramount preoccupation of biometric technology. And to fulfil this task, the body is turned into an informational object, a readable text (or rather *palimpsest*) for statistical (re)measurements and data storage. At the same time, however, it should be borne in mind that biometrics is *not* simply about 'verifying' a pre-given or pre-registered identity by measuring the sameness of body (one-to-one match). If that were the case, biometrics would then be 'an innocent technological practice that only in a rather trivial sense is concerned with personal identity' (van der Ploeg, 1999b: 40). Rather, biometrics is also about 'identifying' and 'distinguishing' one person from another, not just in a technical sense (one-to-many match), but in a much broader sense whereby technology itself becomes actively involved in creating and establishing identities. Homi Bhabha (1994: 64) reminds us that

> The question of identification is never the affirmation of a pre-given identity, never a self-fulfilled prophecy – it is always the production of an 'image' of identity and the transformation of the subject in assuming that image.

Balibar (1995: 187), in fact, goes to the extent to suggest that

> In reality there are no identities, only identifications: either with the institution itself, or with other subjects by the intermediary of the institution. Or, if one prefers, identities are only the ideal goal of processes of identification, their point of honor, of certainty or uncertainty of their consciousness, thus their imaginary referent.

This, to be sure, is true of the case of biometric identification. At first glance, and partially at least, Balibar's proposition, that there is no identity, but only identification, seems to reverberate closely with the biometric project. For the latter appears to be, more often than not, driven by the quest for identification/authentication rather than 'identity itself' (see also Muller, 2004). Not that the *ideal* of identity completely evaporates in the midst of biometric processes. Rather, identity and identification seem to be implicated in a relationship of interdependency wherein identification functions as a process of construction through which forms (or images, to use Bhabha's term) of identity come into being,[9] while the (re)establishment of identity remains as that which provides the impetus and justification for the *raison d'être* of identification techniques.

As we have seen in the previous chapter, and through the examples of Eurodac and the Application Registration Card, biometric procedures contribute to the *establishment* of identity rather than merely the *verification* of a pre-given one; that is to say, biometrics is 'not merely descriptive, but *constitutive* of identity' (van der Ploeg, 2009: 88). In fact, what lies at the heart of biometric procedures is the institutional and governmental will to *bypass* other more 'organic' methods of verifying identity (including the story that is told by the applicant, language analysis, psychological assessment, etc.) insofar as these methods are perceived as contingent and insufficient: 'If a person shows up with nothing with them but the clothes they wear and the story they offer, it would, of course, be a golden solution to be able to *produce* from the person's body an identity' (van der Ploeg, 1999a: 300, emphasis added). The following snapshot is a case in point:

> Bango carries no passport, shouts "asylum!", and claims to come from Sierra Leone. The immigration service interrogates him and lets him take a 'Sierra Leone exam'. Which ethnic group lives in the North-East? What is the name of the largest shopping street in Freetown? Bango fails his exam, the immigration service rejects his application for asylum. He appeals and keeps claiming to come from Sierra Leone. This, like coming from Angola or Afghanistan, would entitle him to a temporary residence permit. The judge does not believe his story.
>
> (in ibid.: 297)

So, through the paraphernalia of technical procedures such as those of biometric technology, identity 'becomes that which results from these efforts' (ibid.: 300) – an identity that is at once independent of the story of the person, and yet 'undeniably belonging to that person' (ibid.). Circling back to the issue of the biometric body, we may suggest that in certain contexts, as in the problem field of asylum, the body becomes more than a mere object of measurement and scanning, but a subject *par excellence* from which identity emerges – at times, against the will and beyond the choice of the person. Through biometric identification, the 'raw' instant truth that is distilled from the body during the procedure of enrolment is processed further and turned into a 'refined' truth. This refined truth forms the basis for processes of profiling, sorting and categorisation. It also tells a story, the story about 'how many times an individual has crossed a border or attempted to enter a country illegally, about an individual's DNA profile [...] how old he or she really is' (Aas, 2006: 153). Ostensibly, however, this story hardly relates to 'personal knowledge about people and the *causes* of their actions' (ibid., my italics) insofar as it is a story told from the *one-dimensional* perspective of the machine/the operator. It excludes ipseity. This constitutes perhaps both the failure and the dream of biometrics: failure to/dream of access(ing) the nexus of the whoness of the person where intentions, actions, beliefs, values, experiences and, indeed, 'resistance' reside.

Deleuze (1992) is undoubtedly right in suggesting that, in control society, individuals are turned into 'dividuals': bits and numbers scattered around databases and identified by their pins, profiles, credit scoring and so on, rather than their subjectivities (see also Rose, 1999: 234). Aas (2006: 155) makes a similar argument in the following way: '[t]echnological systems no longer address persons as "whole persons" with a coherent, situated self and a biography, but rather make decisions on the bases of singular signs, such as a fingerprint'. This *dividuation* has, indeed, much resonance with biometric technology. In fact, biometrics goes a step further. It facilitates the reassembling of those bodily bits in a movement that can be imagined as electronic suturing whereby identities are stitched up or designed from scratch in order to imbue those profiles with a life of their own (a life that might even negate, wipe out or, at least, momentarily override the 'lived life' of the person under scrutiny, as it is often the case with asylum seekers). And through this movement, resubjectification can take place and individuality can (re)emerge again, producing what might be called a 'recombinant identity'. It is a quasi-artificial, but by no means disembodied, identity generated through the combining of various data and whose actualisation and institutionalisation certainly interfere with and affect the life course and the personal 'story-to-come'. Some aspects of this notion of 'recombinant identity' resemble Haggerty and Ericson's (2000) notion of 'data doubles' by which they refer to the process of breaking down and abstracting the body into a series of data. Nevertheless, there is a crucial difference between the two. Whereas 'data doubles' mainly designate a 'decorporealized body' and an 'abstract' type of individuality comprised of 'pure virtuality' and 'pure information' (Haggerty and Ericson, 2000: 611–14), 'recombinant identity', on the other hand, connotes mainly the 'actuality' of re-individuation, that is to say, the terminal point at which data recombine into an identity in the 'concrete', 'corporeal' and 'material' sense. And, never, at any stage, does the notion of recombinant identity consider the body as 'purely' virtual, decorporealised, disembodied or immaterial.

For these reasons, one might justly express a reluctance towards the suggestion that biometrics is merely about the 'what' aspect of a person, or that it is simply concerned with the idem element of identity. For, although biometric technology does not seem to be making much attempt to access whoness and ipseity (or perhaps *cannot* do so[10]), it does, nevertheless, flirt with them, and at times, forcibly so. Not so much in terms of its identificatory *objectives* that remain fixated on what can be distilled from bodily particularities, and even less so in terms of the *specificity* of its technical procedures (assuming here Heidegger's proposition that 'the essence of technology is nothing technological'). But certainly in terms of its wide-reaching *outcomes*, and especially, in terms of the way in which it ends up partaking of processes and practices that *impose* certain recombinant identities and thereby affect the embodied existence of the person. This is particularly true of marginalised groups, such as immigrants and asylum seekers, whose life stories are continuously being shaped by their Sisyphean interactions with bureaucratic institutions and the forms of whatness that are often imposed upon them as a result of such interactions. As Bauman (2004: 13) rightly argues,

> 'Identities' float in the air, some of one's own choice but others inflated and launched by those around, and one needs to be constantly on the alert to defend the first against the second; there is a heightened likelihood of misunderstanding, and the outcome of the negotiation forever hangs in the balance.

Here, indeed, lies in one of the (bio)ethical challenges of biometric technology. The challenge to defend ispe-identity, that self-attesting dimension of who someone is, from institutional impositions – especially when those who 'inflate' and 'launch' enforced forms of identity are chiefly the politicians, policy makers, technical experts, industry representatives, and other 'administrators without responsibility' to put it in Arendtian terms, who, in the name of security and public interest, gather together to *decide* which identities are worthy of the name and which identities are disposable, implausible, if not even, exterminatable. In this sense, the challenge is certainly that of making room, no matter how small and humble it is, for narrative, for self-attestation, for ipseity, for stories, in order to *interrupt* the 'substantialist' formulations of identity and their accompanying myth of communal 'essence' and foundational 'origins' (see Nancy, 1991).[11] It is the challenge of replacing the 'at distance' of the technological[12] with the 'up close'[13] of the personal, of 'listening' to the body instead of 'reading' off the body, and of confronting the technicist and stodgy zeal for sameness with the delicate and affective touch of whoness.

Narrative bioethics of biometrics

Doubtless, the dissolving of the question of 'who' into the question of 'what', of which Arendt, Cavarero and Ricoeur speak, has had a profound and significant impact on the field of ethics itself. More specifically, it has certainly been instrumental to the inauguration and upholding of the universalistic and foundational principles upon which the mainstream styles of ethics have been calibrated, and in defining *in advance* what 'counts' and 'qualifies' as an ethical issue in the first place. This is so inasmuch as the focus on the what instead of the who, on the abstract *universality* of Man instead of the fleshy and situated *singularity* of the person, has led to the foregrounding of rational, meta-theoretical, top-down and rights-based forms of ethics, and thereby disregarding contextual, situational and emotive approaches (see Haimes, 2002; Hedgecoe, 2004). Of course, the reductionist principalism and utilitarianism of mainstream ethics has not remained unchallenged. In fact, the last few decades have witnessed burgeoning attempts, within various fields and disciplines, to rethink ethics beyond the narrow contours of moral theory and outside the abstract ambit of generic principalism. This has particularly, but by no means exclusively, been the case vis-à-vis the fields of biomedicine and biotechnology whereby the interface between life/body and ethics is staged most explicitly. One notable example of such attempts has been the growing adoption of narrative approaches within the interdisciplinary realm of bioethics.

   Narrative bioethics, as the name suggests, can be described as a form of ethics, which takes the notion of narrative as both the ground and the object of ethical reflection and moral justification while addressing issues surrounding life and its technologies. Echoing Rita Charon, a physician and a literary scholar, Arras (1997: 70) describes this ethics as 'a mode of moral analysis that is attentive to and critically reflective about the narrative elements of our experience'. The import of this ethical style into the biomedical and biotechnological field, for instance, has been productively used to challenge the authority of traditional medical ethics by bringing to the fore the complexities and nuances of patients' stories, and to enhance physicians' *responsiveness* towards their patients' suffering instead of taking refuge in the guise of professionalism, objectivity and medical detachment (Brody, 1997; Montello, 1997).

Much of the conceptual underpinning of narrative bioethics is informed by the work of hermeneutics wherein a special emphasis is placed upon the importance of interpretation as an ethical activity and a means of moral evaluation. The practical advantage of hermeneutics, Stepnisky (2007: 198) explains, lies in the way in which it allows us to 'understand the interpretive process that unfolds in the encounter between self and other'. It also lies in its ability to provide a valuable means for countering, or at least complementing, those positions which 'too quickly leave behind the problem of selfhood, and the more intimate forms of self-interpretation' (ibid.: 199). Importantly, such a process of interpretation is by no means complete nor does it strive to achieve a stable meaning. Rather, it remains open to incessant reinterpretation and expandability. 'This emphasis on the ongoing interpretability of things', according to Stepnisky, 'should ease any fears that hermeneutics, despite its appeal to self-understanding, seeks a stable autonomous self' (ibid.: 198).

At the methodological level, there are many ways in which narrative can be used to critically address the field of bioethics. Nelson (1997: x), for example, cites five approaches of doing so: *reading* stories, *telling* stories, *comparing* stories, *literary analysis* and *invoking* stories. In each of these methods, narrative is regarded as a heuristic device for cultivating ethical imagination and enriching the moral landscape. It is not the place here to discuss in great depth and detail the particularities, advantages and limitations of each of these techniques. Suffice, for the purpose of the present chapter to look at how a narrative approach can help us rethink the bioethics of biometrics, specifically in relation to the case of asylum and along the lines of what has been discussed hitherto with regard to Cavarero's and Ricoeur's aforementioned arguments.

As stated at the outset, recent debates on the ethical implications of biometric technology have been largely dominated by rights-centric discourses and permeated with a series of blanket terms such as those of privacy, dignity and liberty. They, therefore, remain implicated within the very same universalistic approaches to ethics, and confined to the very same reductionist definitions of identity in which the question of 'who' is all too often diluted into the question of 'what'. Given its strong engagement with the issue of whoness, one may hope that a narrative approach to bioethics can act as an antidote to practices, including those of biometric identification, that seek to 'simplify' and 'fix' the notion of identity and deprive selfhood of its story. This, however, should be considered neither as a methodological bid to overtake mainstream approaches to the ethics of biometrics nor as a means of erecting a divide between them. Instead, the inclusion of a narrative fibre into the principal dietary regimes of those approaches may help rendering them more mindful and, indeed, 'bodyful'[14] of the ethical force residing in the person's *petit récit* insofar as '[n]arrative provides us with a rich tapestry of fact, situation, and character on which our moral judgements operate' (Arras, 1997: 82).

Returning to the issue of asylum, it is often argued that one major challenge facing immigration authorities and the like is the management of individuals who possess no documents of identity: 'police officers are particularly frustrated over all the identityless asylum seekers of various ethnic origins which are totally out of control' (*Aftenposten* in Aas, 2006: 147). This notion of 'identityless asylum seekers', as Aas explains, is underlined by the assumption that 'identity is something detached from one's self' and that these asylum seekers 'do not have the kind of identity required by state bureaucracy: a stable, objective, unambiguous and thing-like identity'. In fact, this notion represents an instance of what Ricoeur (1992: 149) calls 'man without properties'[15] who 'becomes ultimately nonidentifiable in a world [...] of qualities (or properties) without men'. However, contra the anxiety-inducing formulations of immigration authorities, nonidentifiability and lack of properties (documents of identity in our case), in the Ricoeurian sense, are not necessarily tantamount to a source of frustration and threat. They rather represent 'moments of extreme destitution' whereby 'the empty response to the Question "Who am I?" [i.e. "I am *no one* for I possess no attributes, no papers"] refers not to nullity but to the nakedness of the question itself' (ibid.: 166–7). They, therefore, constitute a remarkable opportunity[16] for 'exposing selfhood by taking away the support of sameness' (ibid.: 149).

In this respect, whereas the practice of biometric identification covers up the nakedness of the question 'who?' by giving it back the flimsy veil of sameness, a narrative bioethics seeks to maintain and perpetuate this state of nakedness by reintroducing the character of ipseity at the heart of identity. In so doing, this ethics places 'the demand for recognition of the *ipse*' (ibid.: 96) while revealing the fact that 'not only [...] *who* appears to us is shown to be unique in corporeal form and sound of voice [elements that can be captured through biometric technology], but that this *who* also already comes to us perceptibly as a narratable self with a unique story' (Cavarero, 2000: 34). As such, this ethics is primarily an ethics of *responsibility* towards the story. It is an ethics of listening and 'suffering-with' (Ricoeur, 1992: 190), an ethics of sympathy that is 'distinct from simple pity, in which the self is secretly pleased to know it has been spared' (ibid.: 191).

For Ricoeur, following the line of the Arendtian thesis, the question 'who?' is inextricably linked to the notion of action, and action is precisely that which calls for narration as a means of saving itself from the abyss of oblivion and saving 'the reciprocal exhibitions of the actors from the fragile actuality of the present to which they belong' (Cavarero, 2000: 26). To this notion of action, Ricoeur (1992: 18) also adds the notion of suffering, linking narrative identity and its ethical dimension to 'the broader concept of the *acting and suffering* individual'. As Marta (1997: 204) puts it: '[t]he "one who acts" is also the "one who suffers" – joy, pain, sorrow, triumph, defeat. The "one who acts," who suffers, bears the ethical and moral responsibility of his or her actions in relation to another and to others'.

It goes without saying that fleeing prosecution and danger is perhaps one of the most powerful examples of acting and suffering[17]: '[t]o flee is to produce the real, to create life, to find a weapon', according to Deleuze (in Nyers, 2003: 1069). Small wonder, then, the issue of asylum has become one of the toughest tests for both politics and ethics, and a strong reminder of the limitations that inhere to the institutionally imposed identity ascriptions. Seen from the vantage point of narrative bioethics, the identity of the person seeking asylum cannot be dissociated from her embodied experience nor can her singularity be extracted merely from the collection of body-data. Rather, the identity of the person becomes the identity of the story itself, an identity recounted and exposed in the presence of another, namely the immigration officer. This scene of exposition and narrativity constitutes the ethical plane of relationality upon which ipseity reveals itself, and with it, the role played by *feelings*. 'For it is indeed feelings that are revealed in the self by the other's suffering, as well as by the moral injunction coming from the other, feelings spontaneously directed toward others' (Ricoeur, 1992: 192).

Therefore, to replace the story with the template, to replace listening with scanning, is akin to amputating the possibility of 'feeling with' (Marta, 1997: 206) and castrating the opportunity of exposing selfhood and uniqueness. Moreover, not only does the paradigm of biometric identification trample upon the ipseity of the person seeking asylum[18] but also upon the ipseity of the person assuming the role of the immigration officer. For it reduces her to the mere executor of a 'power without narrative' (Simon in Aas, 2006: 150) who, even in the case of *giving* refuge to the other, falls short of taking account of the other's singularity and whoness precisely because of the absence of listening and feeling with. In so doing, biometric identification ends up segregating between the person 'acting' as an immigration officer and the person seeking asylum, while confining each to the narrow and dichotomised roles of the giver of refuge (who is 'able to act') and the seeker of asylum (whose capacity to act has been reduced to the sole and silent status of *receiving*).[19] This in turn takes solicitude and sympathy out of the encounter, leaving instead a sterile and simplistic, if not even patronising, sense of charity and benevolence. 'In true sympathy', Ricoeur (1992: 191) writes,

> the self, whose power of acting is at the start greater than that of its other, finds itself affected by all that the suffering other offers to it in return. For from the suffering other there comes a giving that is no longer drawn from the power of acting and existing but precisely from weakness itself. This is perhaps the supreme test of solicitude, when unequal power finds compensation in an authentic reciprocity in exchange, which, in the hour of agony, finds refuge in the shared whisper of voices or the feeble embrace of clasped hands.

Reflections on the limitations of narrative ethics

Despite the above-discussed strengths and advantages of narrative ethics vis-à-vis biometrics, this approach is not without its limitations. For one thing, such an approach cannot take us as far as to fully understand the ways in which identity, security and asylum emerge as 'problem spaces' in the first place, or how biometric technology is activated as a 'technique of governance' and an apparatus of normalisation. Another limitation lies in the fact that power dynamics as well as institutional contexts are not always factored into the narrative perspective on identity and its securitisation. For instance, the relationship between an immigration officer and an asylum seeker is by no means a neutral one. It is rather imbued with a specific kind of power and framed within a specific institutional context, both of which have an undeniable and considerable bearing on the mode of address and on the interlocutory scene within which the story is recounted. Put simply, the inquisitorial tone and the probing frame by which the immigration officer asks the question 'who are you?' already set the stage for and the limits of what can be recounted about oneself during the process of seeking asylum. In more general terms, many of these issues have been famously taken up by Foucault, especially in his consideration of the notion of truth-telling and the formation of the self. The Foucauldian governmentality and subject-formation thesis, in this sense, can help us understand the discursive constructions of identity and the ways in which biometric technology straddles the domain of power and knowledge. This approach, however, remains limited in scope as well, precisely because of its lack of engagement with the minutiae of personal experience and the narratable aspect of selfhood.

The encounter of the narrativity thesis with the Foucauldian theory of subject-constitution is also what animates some of the discussions in Butler's *Giving an Account of Oneself* (2005). Here, Butler productively labours at the intersection between the different theories and philosophies of the self providing another useful lens through which one can trace and juxtapose some of the above limitations. Central to her argument is the idea that the very possibility of narrating oneself is *dependent* on social norms and circumscribed by the structure of address involving others. What fellows from this fundamental and irreducible dependency, according to Butler, is the impossibility of giving a *full* account of oneself and providing a *definitive* life-story insofar as

> the very terms by which we give an account, by which we make ourselves intelligible to ourselves and to others, are not of our making. They are social in character, and they establish social norms, a domain of unfreedom and substitutability within which our 'singular' stories are told.
>
> (Butler, 2005: 21)

As such, stories do not become *recognisable* stories by simply being told. They have to go through the sieve of many social and linguistic conventions to be deemed worthy of recognition.[20] Nor is the 'I' in a full and exclusive possession of its own story. So, in addition to the fact that 'narrating, like saying, calls for an ear, a power to hear, a reception' (Ricoeur, 2005: 251) as well as exposure and co-appearance (Cavarero, 2000), narrating is also irredeemably at the mercy of norms. And whether the story moves us to tears or cripples us with laughter, norms remain indifferent for they are impersonal and do not coincide with the temporality of one's life. 'Discourse is not life; its time is not yours', according to Foucault (in Butler, 2005: 36). By subscribing to this Foucauldian stance, Butler introduces an important caveat that challenges Cavarero's take on narrativity: to the extent that one's account is reliant on norms that happen to exceed oneself, any attempt to give a coherent, authoritative and full-fledged story is bound to be *interrupted* by the time of the discourse, by that very language one deploys as a vehicle for giving an account of oneself. For Butler, this means that singularity itself is subject(ed) to being contested by the temporality of norms.

While this is certainly a valid argument, one could, however, equally argue that the interruption brought about by language as well as the criss-crossing of the temporality of norms with the temporality of life only serve to *reaffirm* singularity, or more specifically, the plurality of singularity. For even if 'I' has to substitute itself to norms in order to tell its story, the way it does so remains singular *through and through*. Each time is a different time and the way the story is told is a singular story in itself. Singularity does not evaporate with reiteration but only consolidates its unrepeatability and strengthens its resistance to being completely dissolved by/into norms. And while exposure is at once a singularising and collectivising experience (ibid.: 34–5), this does not necessarily make singularity any less singular, but only yields a 'singular plural' as Nancy (2000) puts it.[21] So, although norms permeate the very fibre of narrativity, submitting entirely to this Foucauldian position would unduly disavow the nitty-gritty processes by which one 'uses' and 'appropriates' the norms, leading to the foregrounding of an abstract universal subject instead and mutely accepting the icy indifference of norms. I agree though with Butler's view regarding the incomplete and non-definitive character of storytelling: '[t]he "I" can tell neither the story of its own emergence nor the conditions of its own possibility' (Butler, 2005: 37). Nor can it tell the story of its end, except in a speculative and fictitious manner. Completeness and definitiveness are but the necropolis of the story of the 'I'. Yet, such views do not necessarily subordinate Cavarero's theory of storytelling to that of norms nor do they weaken its ethical purchase. Instead they solicit the helping hand of another ethics, one that can handle the necessary, but not-so-comfortable, intercourse between narrativity and norms. Before we say a few words about this ethics, it is worth considering some of the consequences of Butler's postulations with regard to our previous discussions on the issues of asylum, biometrics and narrativity.

In approaching these issues through the lens of Butler's arguments, the initial question that immediately surges to the forefront is to what extent can the story of an asylum seeker *truly* capture her whoness and *fully* reveal her singularity? Clearly, the notion of context can hardly be avoided here. Giving an account of oneself for the *purpose* of gaining the refugee status, and the protection it implies, 'consists of speaking the lines that the institutional interpellation sets in place' (Frank, 1997: 34). This entails the *selection* of facts, recollections and experiences that would qualify the story as a recognisable asylum story, and the use of a specific idiom that would allow the story to fulfil the manifold criteria required for obtaining asylum. Whether in terms of application forms or interviews (which often involve the presence of an interpreter complicating all the more the meditating structure, scene and mode of address), linguistic and institutional norms play a pivotal role. Depending on how they are used and in what circumstances, these norms can either enable or constrain storytelling, rendering the possibility of giving a coherent, consistent, watertight and reliable account a highly contingent enterprise. This is more so the case when the asylum applicant is summoned to undertake more than one interview or fill in more than one application form in order to establish the veracity and validity of her account. Added to that the cases where the person, due to her history of torture and its debilitating effects on the first-person perspective, is unable to construct and articulate an integrated and meaningful life-story that can faithfully and accurately attest to that history and to her embodied ipseity in general: '[some] actual experiences may be too complex, too confusing, too provocative, too shameful, too private, or too common to convey without the help of a "made story" of some kind or other' (Greenspan, 2003: 109). In such contexts, the made story will inevitably be subject to changes, revisions, variations and reinterpretations, despite any attempt to make it otherwise; that is, to turn it into a full-fledged account that is sealed with a permanent stamp of truth and accuracy. As Arthur Frank (in Brody, 1997: 20) argues,

> The 'same' story, retold on different occasions over a span of time, will be heard differently. The self actually engages in change and reformulation by retelling the 'same' story. Thinking with stories thus demands that we attend carefully to how a story is *used* when it is told, how different meanings or shades of meaning are assigned to the story as a result.

Or again

> It's well known that telling and retelling one's past leads to changes, smoothings, enhancements, shifts away from the facts [...] The

implication is plain: the more you recall, retell, narrate yourself, the further you risk moving away from accurate self-understanding, from the truth of your being.

(Strawson, 2004: 447)

Storytelling, in this sense, seems to unfold on a continuously shifting ground and occupy a peculiar and paradoxical space wherein the self is partially concealed (from itself and from others) at the very moment of its own revelation, and narrative is that which testifies to the inability of bearing witness to one's own emergence and constitution rather than to the self-assured capacity to give a full account of oneself. It is as though hide-and-seek is the name of the game that permanently entertains the relationship between storytelling and the truth of one's being. Pitched in this way, one may be tempted to promptly dismiss of narrativity as a method of conveying whoness and housing singularity. For how can a thesis, which is too changeable, fluid, precarious, paradoxical and context-laden, possibly provide an anchoring point for the story of the self, let alone be used as a reliable means of thinking and doing ethics? However, to dismiss of narrativity on these grounds would be too facile a conclusion. In fact, it seems to me that what is at issue here is not so much whoness and singularity per se, but the enduring epistemological and 'technical' questions of truth and validity. The question is not whether storytelling is capable of revealing who one is, but whether this revelation is erupting out of the fountain of truth or emerging from the dungeons of fiction and confabulations. What if the story is not only a 'made story' but also a 'made-*up* story'? What if narrativity is but a futile act of sucking on the 'honeycomb of memory'[22] and risking the sting of the past without any promise or guarantee of finding a *valid* and *working* compass to guide one's decisions (moral or otherwise)?

By raising these questions, we are obviously coming full circle – a move that may well be perceived as a self-defeating detour, since it is in danger of reactivating the all-too-familiar epistemic doubt regarding the story and thereby leaving the room wide open for biometrics to gain a firmer hold on the sphere of identity, to strengthen and capitalise on its truth claims to accuracy and validity. Nevertheless, admitting the limits of the narrativity thesis does not amount to a total defeat. It only emanates a sense of humbleness (unlike the haughtiness of the biometric paradigm) vis-à-vis the general ability of truthfully capturing and divulging whoness and singularity. It is indeed this humbleness (something that Butler herself affirms in her critique of narrativity) that opens up rather than forecloses the horizon of *nonviolent* ethics and preserves rather than destroys the creative dimensions of the different person perspectives. It is this humbleness that makes us aware of 'the fragility of all human communication – its inevitable limits and uncertainty because of its reliance on forms (and, I suppose, beings) that are themselves inherently limited and uncertain' (Greenspan, 2003: 110). It is also this humbleness which reminds us that '[t]hinking with stories means that narrative ethics cannot offer people clear guidelines or principles for making decisions. Instead what is offered is permission to *allow the story to lead in certain directions*' (Frank in Brody, 1997: 20–1). To be fixated on truth and validity is to lose sight of this (ethical) *opportunity*. It is to obstruct the story's lines of flight and to bring the *movement* of decision to a halt (hence the *immobilising* and *limbo-like* character of rigid asylum and immigration policies and technologies). It is not that truth is unimportant. But in the context of storytelling and narrative identity, truth and fiction are inextricably intertwined with no viable possibility of absolute disentanglement. Put simply, fiction is not necessarily devoid of truth nor is truth necessarily non-fictional. As Strawson (2004: 446) argues, '[w]hen Bernard Malamud claims that "all biography is ultimately fiction", simply on the grounds that "there is no life that can be captured wholly, as it was", there is no implication that it must also be ultimately untrue'.

This sense of humbleness in narrativity does not only touch the question of truth, but extends to cover, in a related manner, the notions of definitiveness, completeness and fullness with regard to the life-story. As mentioned earlier, the possibility of giving a full, authoritative and definitive account of oneself is continuously interrupted by the temporality of norms. Death is the only plenitude, the real terminus of every life-story. Because 'I' is *in* time, it is never *on* time. 'I' is always missing an appointment by either being too *late* for the rendezvous with its origin, or too *early* for the rendezvous with its end. Its account is an amputated account made out of prosthetic and phantom narratives. Paradoxically, it is precisely this temporal belatedness or prematurity that injects the 'I' with the possibility of creating itself anew and devising its own stories. Were it not for this *décalage*, 'I' would be capable of neither formation nor narration. In a way, then, before 'I' can stand up with pride and say: 'I know', it has to admit to itself that it does *not* know. Before 'I' can stand up with poise and declare 'I can', it has to come to terms with the fact that it *cannot*. Before 'I' can *stand up* at all, it has to tremble, lose balance and fall. The capacity of the 'I' is, therefore, continuously haunted by its own incapacity. Its potency is constantly threatened by the shadow of its own impotency. Its transparency is often eclipsed by its own opacity. This translates, as we have seen so far and through Butler's critique, into a partial obscurity and a lack of completeness and definitiveness vis-à-vis the life-story, elements that beg for humbleness and fragility (rather than sovereignty and power) as ways of accounting and relating. From here transpire at least two conclusions: one of which has to do with the other ethics, while the other has a direct and practical bearing on the 'everyday' life of the person seeking asylum.

As regards the latter, it concerns the ways in which the non-definitiveness of the story, while representing an intrinsic limitation within the narrativity thesis, may also represent an opportunity. This opportunity is nothing other than the opportunity of saving the story from becoming a snare.[23] Were it not for this non-definitive character, the sealing and authoritative prospect of the 'once and for all' of the story might turn narrative itself into a straightjacket restricting the ebbs and flows of what remains of one's lived life outside of and otherwise than that particular story. In the context of asylum, this becomes a crucial point especially once the refugee status has been granted. Dwelling, in a definitive way, *in* the asylum/refugee story runs the risk of *totalising* identity and fossilising the person into the mode of being a refugee. This, in turn, can have many negative implications not least in terms of hindering the process of genuine (rather than merely functional) inclusion and belonging into the host community, unwittingly encouraging a sense of a disabling and extended over-reliance on the story and on what comes out of it as a bundle of charitable, and in many ways superficial, benefits (e.g. asylum vouchers, which unconstructively strengthen the 'poor me' sentiment), and impeding the person's potential and attempt to reconstruct her life beyond the asylum story and without having to carry indefinitely her refugee status as a badge of identification. There is certainly *more* to the 'refugee' than her refugee story despite the fact that her singular refugee story is an integral *part* of who *and* what she is. That is not to say that the story must be washed away with the detergent of *forgetting*. Forgetting, 'that thief of time' as Ricoeur (2005: 118) refers to it, would be, in this case, akin to committing an act of blasphemy and betrayal towards the pain of the story. What is needed instead is an ethico-political approach, which extends beyond the mere provision of a safe haven to enable the person to develop and explore different ways of *relating* to and *remembering* the story so as to successfully *integrate* its pain into the fabric of her being instead of permanently *identifying* with it.

Undoubtedly, one might wonder, at this stage, if the narrative approach (with its qualities, challenges and visions) towards asylum policy can be amenable to practical application. In a neoliberal culture that is predominantly concerned with security rather than solicitude, with control rather than trust, with power rather than equality, with self-interest rather than care for the other, such an approach may come across as being too theoretical, if not even too unrealistic to be precise. How could narrative ethics possibly pierce through the thick bubble of asylum policy, a policy that seems to be increasingly functioning under the spell of biometric solutions? How could its humbleness, fragility and uncertainty possibly compete with the luring hi-tech veneer of biometrics and its haughty claims to accuracy, truth and objectivity? In their very specificity, these questions are also able to invoke something of a more general dimension, something to do with the hiatus between ethics and (technocratic) politics, which for so long has been the source of many aporias, conflicts and contradictions. While there might not be exact 'ethical' answers to such questions, I do feel however that, if it is to be feasible at all, narrative ethics has to be preceded by, and contribute to, a radical *transformation* at the level of the mental schema that currently governs the landscape of politics and its exclusionist policies of border management, asylum and immigration. Without the necessary shift from the death-producing[24] politics of control to a responsible *and* accountable politics, the narrative approach itself might do more harm than good to the person seeking asylum. For it might risk turning into a *confessionary* apparatus instead of providing a space for solicitude and sympathy. Without this shift in the political imaginary, asking the policy-maker to give up biometric control in favour of narrative ethics would be like asking a vampire to give away her fangs to the dentist. Nevertheless, instead of resorting to cynicism, one can, as a starting point, intervene by demonstrating how such policies do not only fail but also worsen the situations they seek to remedy. It is a matter of heightening policy makers' *awareness*[25] that fighting against unwanted immigration and asylum with technology or otherwise only ends up producing an even more unmanageable chain of problems such as human trafficking, death at the border and exploitation. And this is perhaps the tragedy of contemporary forms of governing; the more problems they seek to solve, the more problems they create. After all

> Migrants and those who facilitate their migration resort to staggering feats of ingenuity, courage and endurance to assert their right to move and to flee [...] The question is how much suffering will be imposed on innocent people, and how much racism will be stoked up [...] before governments finally abandon the effort.
>
> (Hayter, 2000: 152)

From the concatenation of the above reflections, it is clear that, if taken as a stand-alone approach, narrative bioethics would not always be able to single-handedly tackle the manifold challenges pertaining to the field of asylum and biometric identification. This limitation is, in fact, what calls for a well-rounded 'coalitionist ethics' whose approach must be based on the cross-pollination and cross-fertilisation of different, albeit contradictory, theoretical and empirical perspectives and an appreciation of the distinct qualities of each, and whose primary task is to *question* the very *foundations* upon which contemporary political styles of thought and practices are based – aspects of which will be discussed in a later chapter.

Conclusion

We began this chapter by interrogating the ways in which biometrics is about identity and uniqueness in an attempt to uncover some of the bioethical stakes of biometric technology. This interrogation has led us straight into the quagmire of asking what identity is. Drawing upon the work of Cavarero, Schechtman and Ricoeur, we explored some of the variations in the meaning of identity. Emphasis has been placed upon the distinction between the question 'what?' and the question 'who?' through which we examined the interplay between biometric technology and identity. Although, at first glance, biometrics may seem to be mainly concerned with the 'what' aspect of identity, we argued that the 'who' dimension is inevitably implicated as well, especially in the context of asylum. Given the importance of narrative to the question of 'who' and to the notion of uniqueness, we proposed a narrative approach as a means of navigating through the distinctive bioethical implications of biometric technology. Our discussion has shown that, paradoxically, in its pursuit of capturing the singularity of the person, biometrics only ends up obstructing the exposure of singularity precisely because of its amputation of narrative from the sphere of identity. Thus, a pressing bioethical task would be to seek to preserve the narrative dimension of identity which, in the words of Cavarero (2000: 34), constitutes the 'house of uniqueness'.

Our enquiry has also revealed some of the limitations of the narrative approach towards the bioethics of biometrics. We argued that these limitations are mainly the result of the intricate and restrictive relationship between narrativity and the socio-linguistic norms, as well as the institutional contexts and power dynamics within which relations (such as those between the asylum seeker and the immigration officer) are embedded and conducted. But despite all its limitations, narrative will still remain 'an indispensable and ubiquitous feature of the moral land-scape' (Arras, 1997: 68). So, for the time being and in the context of this chapter, let us be content with the conclusion that the moral of the story is perhaps nothing other than *listening* to and *feeling* the story itself.

4
Identity Securitisation and Biometric Citizenship

In the previous two chapters, we looked at some aspects of biopolitical dimensions and bioethical implications of biometric technology and identity systems. Our discussion has been primarily focused on the domain of asylum and on the ways in which biometric technology functions as a means of managing the identities of those who are held within such a domain of power and control, affecting their embodied existence, as a result. In this chapter, we shall shift the attention towards the figure of the 'citizen' in order to explore other aspects of the interplay between biometrics and identity management and how this interplay relates to the ideal and practice of citizenship, by looking at practices that are less exceptional and more routine than those of asylum. As the title of this chapter suggests, security is a key concept that underpins the triad of biometrics, identity and citizenship. And like many other concepts, security too has undergone many transformations in its meaning, use and function. As such, it is worth starting off the discussion by considering some of these transformations. This will also help us pave the way for analysing and understanding what is involved in the process of securitising identity through biometric technology as well as the impact of such a process on the concept and practice of citizenship.

Security

In recent years, the debates over the nature, meaning and significance of security have become the subject of renewed interest and controversy (Collins, 2007: 2; van Munster, 2005a: 1; Williams, 2003: 512). Traditional approaches to security such as the 'realist' perspective, which dominated much of the early literature of security studies, have been undergoing a series of challenges from different fields and disciplines. These challenges were prompted by various socio-political events including the fall of the Berlin Wall, the collapse of the Soviet Union and the myriad of ethnic conflicts (van Munster, 2005a: 2). They grew out of a sense that 'security no longer has a stable referent object, nor names or common set of needs, means, or ways of being' (Burke, 2002: 2). In other words, 'what' should be secured, 'how' it should be secured and 'from what' it should be secured can no longer be contained within or merely understood in terms of the state-centric (neo)realist approach. Instead, they require what is commonly referred to as a 'deepening' and 'broadening' of security conceptualisations, definitions and scope (Collins, 2007).

As a response to this demand, many theorists started shifting towards a more constructivist view whereby security is treated not as an objective condition but as a process of continuous social and rhetorical construction: '[w]ith the help of language theory, we can regard "security" as a speech act. In this usage, security is not of interest as a sign that refers to something more real; the utterance itself is the act. By saying it something is done' (Wæver in Williams, 2003: 513).

This shift from the ontological to the constructed, from the given to the performative in security studies, is best captured through the theory of 'securitisation'. As initially proposed by the Copenhagen School, securitisation is concerned primarily with the study of 'security practices as specific forms of social construction [...] as a particular kind of social accomplishment' (Williams, 2003: 514). The reformulation of security along these lines has allowed the Copenhagen School to, at once, 'broaden' and 'limit' the conception and analysis of security. For, in one sense, by regarding security as a speech act, an almost indefinite number of threats and referent objects can be subsumed under its rubric. At the same time, however, the securitisation theory delineates certain limitations as to what can be considered as a 'securitising speech act' (Emmers, 2007: 109; van Munster, 2005a: 3; Williams, 2003: 513). As Williams (2003: 513) explains, 'securitization has a specific structure which in practice limits the theoretically unlimited nature of "security"' and treat the latter 'as a phenomenon that is concretely indeterminate and yet formally specific' (ibid.: 516). What distinguishes a securitising act from other forms of speech act is a series of characteristics consisting of three essential elements. First, the issue at hand is discursively represented as an 'existential threat' to security. In order for this process to be effective and successful, the securitising act has to fulfil certain linguistic and social conditions, and convince the relevant audience of the existence and imminence of threat. Second, and depending on the success of the securitising act, the issue is presented as a matter of supreme priority and urgency, requiring the use of 'extraordinary means' and 'exceptional measures'. Third, the securitisation of an issue justifies the breaking free of 'normal' democratic procedures and the 'curbing of civil liberties in the name of security' (Emmers, 2007: 115; see also van Munster, 2005a: 3; Williams, 2003: 514).

In this sense, Williams (2003: 515–6) and van Munster (2005a: 3–4) argue that securitisation theory is heavily influenced by the Schmittian conceptual framework vis-à-vis the notion of 'the political'. For Schmitt, the core of the political is not to be found in the 'issues themselves', but in a particular way of framing and approaching them; just as in the theory of securitisation, the phenomenon of security lies not in the security issue itself or in its intrinsic nature, but in the ways in which it is constructed, presented and accepted as an existential threat.

The notion of exception is, therefore, another aspect where Schmitt's understating of the political and securitisation theory's approach to security converge. In both frameworks, there is an emphasis on the binary groupings between friend/referent objects and enemy/threat, and on the decisionist/performative authority, which determines whether there is to be a case of unanticipated emergency and on what must be done to eliminate it. For Schmitt, the dividing political act rests upon the notion of sovereignty, that is, the ability to decide on the exception. At the same time, the effectiveness of the sovereign decision itself rests upon that dividing political act between friend and enemy in as much as '[f]riendship and enmity provide the foundational structure of allegiance, of solidarity, that underpin the capacity of effective decision' (Williams, 2003: 517). And, in terms of securitisation, '[the] act of decision is both the "primary reality" of securitization and an expression of the existence (in cases of successful securitization), non existence (in cases of failure), or calling into being (creative mobilization) of "political" groupings that feel so intensely about a given issue' (ibid.: 518). Again, it must be stressed that in securitisation theory and unlike in the Schmittian framework, this act of decision is reducible neither to the notion of sovereignty nor to the survival of the state but extends to other areas and concerns. Among these is the notion of identity.

Identity is, indeed, one of the key points of interest and analysis in securitisation theory. It even serves as a demarcating concept between 'state security' and 'societal security'. Wæver (in Williams, 2003: 519) argues that '[s]tate security has sovereignty as its ultimate criterion, and societal security has identity. Both usages imply survival. A state that loses its sovereignty does not survive as a state; a society that loses its identity fears that it will no longer be able to live as itself'. In such a context, just as the sovereign decision relies on the division between friends and enemies, so too does the securitisation of identity:

> Under the conditions of 'existential threat' [...] to identities, a Schmittian logic of friends and enemies is invoked, and with it a politics of exclusion [...] A successful securitization of an identity involves precisely the capacity to decide on the limits of a given identity, to oppose it to what it is not, to cast this as a relationship of threat or even enmity, and to have this decision and declaration accepted by a relevant group.
>
> (ibid.: 520)

From this perspective, the securitisation of identity is considered as a process by which the flexibility and negotiability of identities are contained and suppressed. It is a way of founding and declaring a collective monolithic identity on the basis of the existential threat to which it is supposedly exposed, and through the intensification of certain affects that contribute to the formation of political and social groupings. We have seen in a previous chapter that the construction of the asylum and immigrant identity undergoes similar discursive processes, serves similar purposes of national and international (re)foundings and feeds into the various divisions of us/them, legitimate/illegitimate, belonging/non-belonging and so on. We have also demonstrated how the notion of exception partakes of the overall contemporary modes of governing borders and circulation through biometric technology. And, we have attempted to show that the exceptionalist measures of security are part of the banoptical dispositif and its diagrams of control. All these propositions are indeed very much in tune with the exceptionalist trope of securitisation discussed above and its interplay with the issue of identity.

It should be noted at this point that, as in the case with the conceptualisation of the banopticon,[1] the 'decisionist' feature that underpins the exceptionalist model does not exhaust all aspects of securitisation nor does the Schmittian-inspired understanding of identity exhaust the different modes of its securitisation. This is because both decisionism and the articulation of identity in terms of friendship and enmity focus almost exclusively upon 'extreme' and 'distinct' moments and events. They, therefore, do not often touch upon the routine practices and the daily processes by which emergency is subtly 'normalised' as a technique of 'government by unease' (Bigo, 2006a: 63). This is why Bigo insists on the need 'to go beyond the debate of the exception as a "moment" of decision or as the opposite of a "norm"' (ibid.: 50). In a similar vain, Williams (2003: 520–1) argues that

> [the] stress on decision clearly raises difficult analytic questions, since to focus too narrowly on the search for singular and distinct acts of securitization might well lead one to misperceive processes through which a situation is gradually being intensified, and thus rendered susceptible to securitization, while remaining short of the actual securitizing decision.

So although everyday practices of securitisation might not have the same dramatic element of intensity that is inherent to the logic of decisionism, they still play an integral part in the overall landscape of contemporary security (van Munster, 2005a: 6). This argument becomes particularly pertinent when one considers how, in security discourses and practices, what is at issue are not only 'existential' threats but also 'potential' ones, i.e. risks.

In this sense, another way of formulating securitisation theory is through the lens of 'risk management'. Unlike the exceptionalist model of securitisation, risk management is not interested in the decisionist approach of dividing the population into friend/enemy groupings. For it is driven by the belief that '[t]oday it is increasingly difficult [...] to name a single unified enemy; rather, there seems to be minor and elusive enemies everywhere' (Hardt and Negri, 2000: 189). As such, risk management is more concerned with identifying, profiling and classifying people according to the level of risk assigned to them in order to detect, reduce and neutralise the perceived danger. Risk management is, therefore, based on pre-emptive mechanisms and preventative techniques in which threat is conceived not so much in terms of its actuality but in terms of its potentiality (its becoming dangerous, its 'real virtuality'); that is, 'on the basis of what one might do rather than apprehending one after the act' (Rose, 1999: 241). So, instead of focusing on exceptional events and decisions, risk management techniques are immanent within all areas of life. They permeate everyday flows, transactions and practices. They are based upon 'a dream of the technocratic control of the accidental by continuous monitoring' (ibid.). In fact, the very notion of risk, as Jennings (2007: 2) puts it, is a 'colonising' concept. It easily and ubiquitously creeps into the everyday with all its mundane administrative, organisational and bureaucratic arrangements and activities.

Not surprisingly, then, identity itself is increasingly being recast in terms of risk and security. In the next section, we shall look at a certain problem field within which the securitisation of identity through biometric technology emerges as a solution to the array of risks believed to be facing identity in contemporary society. This will also help us expand on some of the arguments, made in an earlier chapter, regarding the deployment of biometric technology as a means of managing and governing the *entire* population rather than just specific bodies and exceptional spaces, such as those of asylum as previously discussed.

Securitising identity

> On any list of public concerns, illegal immigration, crime, terrorism and identity fraud would figure towards the top. In each, identity abuse is a crucial component.
>
> (Tony Blair, 2006)

The increasing interest in finding stronger and more reliable means of securitising identity is underpinned by a variety of risk-based reasons and technology-driven explanations. The rising levels of global mobility, the advances in new technologies, the dispersion of information networks, the increasing need to control access to social benefits and entitlements, the changing scene of borders and states are all some of the many factors behind and arguments for the deployment of biometrics and the reconfiguration of the means by which the state connects to its embodied (non)citizens and regulates the flows of their mobility and transactions. In this respect, the emerging identity systems are part of a large-scale direction towards governance (Lyon, 2004: 2) in which the management of the life of the population *through* risk is the primary objective and the securitisation of identity through biometrics is one of its main features.

For instance, the UK's 2006 *Identity Cards Act* outlined a set of purposes for implementing an identity cards system. All of these were subsumed under the heading of 'public interest', which included 'national security', 'prevention or detection of crime', 'enforcement of immigration controls', 'enforcement of prohibitions on unauthorised working or employment' and 'efficient and effective provision of public services' (Home Office, 2006a: 2). Couched in this double rhetoric of public interest and prevention, identity systems are often framed within a certain political and regulatory rationality that partakes of wider and ongoing efforts to socialise security, regulate access and infuse a sense of 'prudentialism', while being 'continually open to the construction of new problems and the marketing of new solutions' (Rose, 1999: 160). To this end, identity systems are often promoted as a kind of panacea to social ills and a solution to the various 'problems' brought about by global mobility, terrorism, technological advancements and so on.

According to the Home Office (2006a: 5), national security and detection of crime involves the prevention of terrorism as well as identity fraud and theft. In fact, the notion of identity fraud and theft is a common and recurring theme that runs across and links between every single concern underlying the rationale of biometric identity systems. As per the above statement by Tony Blair, 'identity abuse' is considered as a crucial element in each of the threats believed to be facing the interests of the public. This argument itself stems from the belief that those in breach of immigration law, those engaging in illegal work and unauthorised employment, those committing acts of crime and terrorism, those who are double-dipping and so on, all rely in one way or another on the relative ease by which one can build a new and false identity, appropriate someone else's identity or gain unauthorised access to personal data and financial information. For instance, '[t]errorists routinely use multiple identities – up to 50 at a time – to hide and confuse. This is something al-Qa'eda train people at their camps to do' (Blair, 2006).

Identity fraud is thus increasingly being framed as both a security and a social problem. It is constructed as a specific kind of risk that pervades a myriad of spaces and activities and whose management requires various strategies and techniques, including the securitisation of identity through biometric technology. The argument of identity fraud, as such, functions as one of the primary vehicles for facilitating the introduction and spread of identity systems and insuring the public acceptance of them. It is, therefore, worth examining how identity fraud is emerging as a problem field within the current governmental landscape, and how the problematisation and criminalisation of this issue serve as a mechanism by which the figure of the citizen, the (in)dividual, the consumer and so on can be (self-)managed, (self-)governed and (self-)responsibilised.

In 2006, identity fraud has officially entered the realm of criminal law in the UK through the enactment of the *Fraud Act 2006*. The Act, which came into force on 15 January 2007, creates new offences relating to fraud that can be committed in three ways: by dishonestly making a false representation (Section 2), by failing to disclose information (Section 3) and by abuse of position (Section 4) (Home Office, 2006a). These offences cover individuals as well as businesses, and include a variety of fraudulent activities ranging from credit card fraud to phishing[2] on the internet. So in comparison to other previous legislative provisions, such as the *Theft Act 1968* and the *Companies Act 1986*, the *Fraud Act 2006* seems to consolidate and expand on those provisions by extending the scope and the meaning of fraud offences. Sub-section (5) of Section 2, for instance, states that 'a representation may be regarded as made if it (or anything implying it) is submitted in any form to any system or device designed to receive, convey or respond to communications (with or without human intervention)' (Home Office, 2006b: 2) – which means that there is no limitation as to *how* a representation must be made or implied to be considered as dishonest. This statement is indicative of the legal elasticity of the Act and the criminalisation of various activities that can be carried out through information and communication technology. The *Fraud Act* is, as such, an attempt to 'keep abreast of emerging technologies and to obviate the need for constant reactive reform' (Savirimuthu and Savirimuthu, 2007: 440). In terms of phishing, for example, the Act seems to facilitate the prosecution of this online activity by 'demanding neither proof of deception nor the obtaining of any property, which were pre-requisites to conviction under the previous legislation' (ibid.: 441).

It is clear, then, how the law is being mobilised to respond to the threat of identity fraud through the enforcement of new civil penalties provisions and the reconfiguration of what is entailed by the notion of fraud. For law is, undoubtedly, an important instrument for 'promoting compliance with accepted standards of behaviour and norms', and for influencing 'the way individuals approach risks' (ibid.: 439, 443). There is, however, an increasing awareness from the part of the government, regulatory bodies, public organisations, private companies and so on, that law cannot single-handedly tackle the 'problem' of identity fraud nor is it a sufficient instrument for 'preventing' its occurrence. In fact, what law seems to do best in the context of the *Fraud Act 2006* is to serve as a '(re)problematising mechanism' by which old problems are given fresh makeover and renewed interest, and 'emerging' ones (especially those relating to the ongoing challenges of the online environment) are made amenable to legal regulations and legislative enforcements. And, what comes part and parcel of this (re)problematising mechanism is a series of 'techniques' and 'technologies' aimed at complementing the law and tackling that which goes beyond its capacity (particularly with regard to the issue of prevention). As Miller and Rose (2008: 15) explain:

> [T]he activity of problematizing is intrinsically linked to devising ways to seek to remedy it. So, if a particular diagnosis or tool appears to fit a particular 'problem', this is because they have been made so

that they fit each other. For to presume to govern seemed to require one to propose techniques to intervene [...] In short, to become governmental, thought had to become technical.

Indeed, in addition to the passing of the *Fraud Act 2006*, other techniques have been put in place to tackle the problem of identity fraud. For instance, the Home Office has produced leaflets and devoted a website for raising awareness vis-à-vis the increasing threat of identity fraud and for providing tips on how to protect one's identity and on how to proceed if one falls victim to identity fraud (see www.identitytheft.org. uk). CIFAS, the UK's fraud prevention service, offers a 'Protective Registration' package for an administration fee of £14.10 through which individuals can protect their names and addresses. This service allows the registered client to constantly monitor her credit history and activity through a warning system that prompts the individual each time her name or address has been used to apply for credit, open a bank account, make an online payment, etc. (see http://www.cifas.org.uk/). Furthermore, private companies, especially financial and banking institutions such as Citibank, Barclays, Lloyds TSB and Capital One have also been actively involved in the preventative fight against identity fraud through their advertising campaigns, protective services and online security tips. For example, 'Citi IdentityMonitor' is a service provided by Citibank for users to track their daily credit activities and gain information about potential fraud. The service also offers access to 'credit education' specialists (see: http://www.identitymonitor.citi.com/).

What is particularly interesting about all these campaigns, be they government-led or privately orchestrated, is the ways in which they are geared towards the notion of the 'consumer' and couched in the language of the 'credit market'. They represent identity fraud as 'an incalculable systemic risk stemming from the productiveness of the market itself – a systematic by-product of an advanced, electronically enabled credit system' (Marron, 2008: 23). Identity, in this style of thought, is considered as a valuable asset that enables the neoliberal actualisation of one's autonomy, freedom and choice within the circuits of consumption: '[y]our identity is a valuable commodity – you need it to function in everyday life. You need evidence of who you are to open bank accounts, obtain credit cards, finance, loans and mortgages, to obtain goods or services, or to claim benefits' (CIFAS, 2007). Therefore, what identity fraud seems to threaten is precisely that 'individualised' ability to consume, 'the entrepreneurial potential as a consumer' (Marron, 2008: 24) and 'the personal freedom through which [consumers] are integrated as subjects and objects of government' (ibid.: 23): 'If your identity is stolen, you may have difficulty getting loans, credit cards or a mortgage until the matter is sorted out' (Home Office, 2008b). Identity fraud, in this regard, amounts to something more than a 'constructed' risk. It is also presented as having a 'realist' dimension, which can affect the individual's (credit) history, her (consumer) identity, her sense of continuity and interrupt the 'ability to create a life for oneself though one's consuming choices' (Marron, 2008: 23–5). It is at once a by-product of neoliberal market activities and a negation of their principles and possibilities.

In addition to this financial and material impact, the ramifications of identity fraud are also framed in terms of their 'emotional' dimension, which can represent a 'harrowing experience' for the victim (CIFAS, 2007). According to Marron (2008: 25), the adverse emotional side effects of identity fraud create a particular form of identity; that of '*being a victim*'. This identity is described as being 'static' and 'unyielding' not only in the sense that it is marked by the denial of the possibility of future choice and the inability of being in control of one's own credit identity. But also in the sense that it is an identity through which the victim experiences a myriad of negative affects including the loss of trust, feeling violated, invaded, distressed, depressed and, in some cases, dysfunctional too. For, 'even if the victim might be materially "in the clear", they are still positioned as being encumbered with a debilitating uncertainty that pervades their life, disrupting their sense of "ontological security"' (ibid.: 26). And, around these affective problems gathers a pool of consumer advisory and advocacy groups, insurance companies, fraud prevention agencies, credit-scoring and credit-reference agencies, whose advice is, more often than not, geared towards 'making individuals culpable risk managers' and prudentialising the population (ibid.: 34).

The configuration of identity and identity fraud in these terms can, thus, be seen as partaking of a specific governmental rationality whereby the responsiblisation of individuals is considered not only as a useful accompaniment to legal regulation, but an integral strategy for its success and for the overall pre-emptive fight against identity fraud. The previously mentioned techniques and strategies are indeed examples of the 'consumer education programmes' (Milne in ibid.: 29), which seek to equip consumers with the necessary skills for exercising individual prudence and minimising the risk of falling victim to identity fraud. In so doing, these techniques enable the shifting of responsibility onto individuals themselves instead of containing it exclusively within the remit of government and institutions. As O'Malley (in Rose, 2000a: 328) argues, 'not only does responsibility for crime-risk-management shift, but co-relatively, the rational subject of risk takes on the capacity to become skilled and knowledgeable about crime prevention and crime risks'. This way, the problem of identity fraud becomes framed in terms of individual knowledge (or lack of it) vis-à-vis the prescribed techniques of risk management. It becomes 'pitched not as one of systemic institutional culpability, but as a lack of awareness on the part of individuals [and] a lack of self-mastery' (Marron, 2008: 29–30).

Undeniably, this way of framing plays a crucial role in imbuing biometric identity systems with a sense of 'personalised' and yet 'universal' legitimacy and necessity:

> it is essential for all of us to be able to lock our identity to ourselves and to protect its integrity. We need a way of doing so that we can trust in, and that can be trusted by others – when applying for a job, travelling abroad, or using business and government services [...] I have always believed in the concept of a national identity scheme.
> (former Home Secretary Jacqui Smith, 2008)

Such discursive formulations render biometric identity systems as a 'technology of subjectivity' through which individuals can exercise their autonomy and freedom and securitise their identities against the risk of fraud. In this regard, biometric technology and identity schemes have been marketed as an important component of the 'knowledge toolkit' required for managing and protecting one's identity, as a means by which individuals can actualise their *savoir-faire* and optimise their *savoir-être* within the circuits of consumption. They thus represent a vivid example of the current strategies of 'governing through risk' and inculcating the ethos of neoliberalism that are encapsulated in the notions of self-responsibility, self-mastery, self-monitoring, autonomy and so on (see Rose, 1999, 2000a). They are an expression of 'a deterministic attempt to develop methods of "pre-crime" control' (Mythen and Walklate, 2006: 389) in order to deal with the inherent uncertainty of contemporary risks such as the risk of identity fraud.

At another layer, the logic of biometric identity systems can also be regarded as taking the form of a precautionary principle in the sense that every identity is treated as a suspect identity until proven otherwise through biometric identification. The precautionary principle, as van Assett and Vos (2006: 314) explain, is an important concept for addressing situations whereby 'uncertainty and risk intermingle'. It is a paradigm of risk management that takes action 'not on the basis of what we know, but on the basis of what we do not know' (Gulberg, in Aradau and van Munster, 2005: 11). It therefore activates proactive technologies of preventions, such as those of biometrics and identity systems, in an attempt to respond to risks that are characterised by their unpredictability, uncontainability, contingency and by their challenging nature vis-à-vis the calculability and control of knowledge.

Identity-related crimes, as mentioned before, are fundamentally incalculable and irreducibly uncertain. They feed upon the disembedded, automated and decentralised features of information exchange and credit market processes. They represent a 'phenomenon which exists as an active agent in its own right, aleatorically and malevolently striking unsuspecting, exposed consumers' (Marron, 2008: 34). Interestingly, the perpetrators of identity-related crimes are often considered in terms of their 'disembodied' nature, as being irredeemably elusive and notoriously hard to identify: 'according to one survey, nearly three-quarters of victims had their identity stolen by an individual who was unknown to them (FTC, 2003a: 28) while, within the criminological literature, only Allison et al.'s (2005) exploratory study in Florida has made any attempt to uncover who the offenders might be' (ibid.: 29). What this amounts to at the level of the governmental in general, and in terms of the securitisation of identity in particular, is a sense of perpetual unease regarding the nature of current threats and the impossibility of knowing the identity and the location of the 'enemy' (within) (Bigo et al., 2007). Consequently, governments, institutions, companies, individuals are all becoming acutely aware of the necessity to take 'precaution' against the manifold and fluid risks facing (consumer) identity, a process that requires enlarging the scope of not only the techniques by which risks can be managed, but also the categories under which suspect and 'potentially' dangerous groups can be subsumed. This means that the surveillance and control of the 'hard core' is no longer enough. 'The problem then becomes the criminality and potential criminality of the "soft core", in short the rest of the population' (Norris, 2007: 150). So while biometric identity systems are emerging as a favourite precautionary device for protecting identity and securitising everyday activities, they are at the same time turning every identity into a potential suspect/victim identity by default. What transpires from framing identity as being at risk is a reconfiguration of the relationship between the state and its subjects, and with it, a reconfiguration of the meaning and function of citizenship. In the next section we shall ask what kind of citizenship is the 'biometric citizenship'.

Reconfiguring citizenship through biometric technology

> The introduction of a national identity system will herald a significant shift in Britain's social and economic environment [...]
> For better or worse, the relationship between the individual and the State will change.
>
> (LSE, 2005: 4)

> What is at stake here is nothing less than the new normal bio-political relationship between citizens and the state. This relation no longer has anything to do with free and active participation in the public sphere, but concerns the enrolment and the filing away of the most private and incommunicable aspect of subjectivity: I mean the body's biological life.
>
> (Agamben, 2004)

Traditionally, the idea of the citizen has been one of the most fundamental and, at the same time, problematic premises of Western political thought. Its history is a history of accomplishment and struggle, a history of emancipation and conflict, a history of aporia and paradox. And it is so, precisely because of the very essence and nature of the ideal of citizenship itself. For while citizenship has functioned, since its very beginning(s), as the framework for and the embodiment *par excellence* of claims to membership, rights, freedom and so on, it also remained unavoidably and inextricably linked to acts of exclusion, inequality, oppression and, if not even, violence as well. Clearly, the idea of the citizen has gone through a number of transformations and updates so much that it has developed an almost incurable dependency on having qualifying prefixes for its own definition (e.g. 'cosmopolitan' citizenship, 'postnational' citizenship, 'world' citizenship, 'digital' citizenship and 'biological' citizenship). The majority of these prefixes often run the risk of being (mis)labelled as mere tautological ornaments. Yet, they remain as valid epitomes of the inherently multi-layered nature of the notion of the citizen, and relevant instances of what some have termed 'thin' conceptions of citizenship, which go beyond the 'thick' agency of state citizenship (see Faulks, 2000; Nyers, 2004). Prefixes are, therefore, a handy mechanism for attending to the thinness/thickness aspects of the tectonics of citizenship, and a useful reminder of the necessity to take into account the question of context and the specificity to which it gestures. In what follows, then, we shall venture into pinning the prefix 'biometric' to citizenship and explore what is entailed within such a compound.

The opening epigraphs of this section invite us to reflect on the ways in which biometric technology and identity cards are symptomatic and constitutive of the ongoing mutations that are taking place within the emergent forms and practices of citizenship. These mutations are unfolding in a myriad of problem fields and socio-economic sites, and 'crystallized in an ever-shifting landscape shaped by the flows of markets, technologies, and populations' (Ong, 2006: 499). In the previous chapters, we started addressing some aspects of this shifting landscape by looking at various interrelated examples such as border management, immigration and asylum policy and by examining how biometrics and identity systems are mobilised as a means of managing and securitising these problem spaces. The present section is an attempt to build upon and extend our preceding discussions by zooming further into the kind of rationalities and techniques, which underpin and guide the government of these sites all the while revealing how such modes of governing play out in the domain of contemporary citizenship. The aim is to understand some of the complex and paradoxical features of the changing relationship between the state and its (non)subjects, between subjects and institutions and between subjects and themselves, this, through the examination of some empirical examples relating to biometrics and other related techniques, and through the conceptual lens of the governmentality thesis.

Biometric citizenship as neoliberal citizenship

According to Nikolas Rose (2000a: 324), '[a] whole range of new technologies – "technologies of freedom" – have been invented that seek to govern "at a distance" through, not in spite of, the autonomous choices of relatively independent entities'. Rose's succinct assertion points out to one of the core aspects that characterises the rationality of the neoliberal paradigm of governmentality, that is to say, the art of 'governing *through* freedom'. This modality of governing takes, as its premise, the logic of individual 'autonomisation' and 'responsibilisation' superseding some of the principles of earlier forms of liberal government such as those of welfarism and its 'culture of dependency' (Miller and Rose, 2008: 79). The overarching objective of neoliberalism is to organise individual, group and institutional activities around market-based ethos and active entrepreneurship by way of reactivating the capacities of free individuals while minimising interference from the state, and endorsing processes of marketisation and tehnologisation. In terms of citizenship, this translates into a shift from the ideals of social responsibility and collective solidarity as being the basis for and the currency of membership towards a reconfiguration of citizenship in terms of choice, freedom and the ability to be an active entrepreneur of oneself. As Miller and Rose (2008: 48, 82) argue,

> [t]his citizenship is to be manifested not in the receipt of public largesse, but in the energetic pursuit of personal fulfilment and the incessant calculations that are to enable this to be achieved. [It is] to be active and individualistic rather than passive and dependent.

Much of the emerging practices and reformulations of citizenship are increasingly driven by and imbued with these neoliberal principles of freedom, choice and active entrepreneurship. They are less informed by the traditional norms of the democratic society and the political rights of membership in a collective body, and more concerned with the entrenchment of transnational capitalist criteria of access and entitlements in order to reinforce business-driven models of governing and enable the global and free flow of market processes (see also Côté-Boucher, 2008; Cowen and Gilbert, 2008; Ong, 2006; Sparke, 2006). To understand the ramifications and transformative implications of neoliberalism on the concept and practice of citizenship, it is necessary to examine some concrete examples pertaining to the rationalities, mechanisms and technologies by which neoliberal forms of citizenship are rendered imaginable, thinkable and operable. The case of identity cards and the deployment of biometrics in the field of identity, borders and immigration management provide us with just such sites and examples for empirical investigation.

To proceed, we shall examine how the neoliberal rationality of 'governing through freedom' is instantiated within various identity related programmes. As a starting point, it is worth considering, briefly at least, some aspects of the changing notion of the 'state' itself under the regime of neoliberalism. It has long been the contention of many theorists of governmentality that the figure of the state can no longer be viewed as a monolithic and unified actor with an executive power over society, but rather as 'a specific way in which the problem of government is discursively codified, a way of dividing a "political sphere", with its particular characteristics of rule, from other "non-political spheres" to which it must be related' (Miller and Rose, 2008: 56). They follow Foucault's proposition that '[i]nstead of turning the distinction between the state and civil society into an historical universal enabling us to examine every concrete system, we may try to see in it a form of schematization peculiar to a particular technology of government' (Foucault, 2008

[1979]: 319). This view, however, does not amount to sounding the death knell of *the* state as such, but to recognising its dispersed, networked and polymorphous function, which operates on an increasingly shifting terrain and through a myriad of organisational activities and governmental practices. And within the neoliberal modality of government, the image of the state that is often conjured up is that of the 'enabling', 'facilitating' and 'animator' state as opposed to that of the 'cold monster' (Miller and Rose, 2008: 54; Rose, 2001: 6). Such an image is captured clearly through the former Home Secretary Jacqui Smith's (2008) speech on the identity cards scheme. She states:

> Rather than thinking of the state as an opponent of our liberties, set on thwarting our personal ambitions, in this context the role of government agencies is to defend our interests, to offer reassurance and trust, and to working in the most effective way possible to ease and enable our lives. This is the argument that supports the principle of the national identity scheme.

Smith's argument is an expression of a twofold rationality. On the one hand, it indicates a rationality by which the state itself is 'degovernmentalised' (its responsibilities are decentred and distributed across different entities) and 'regovernmentalised' through the objectives and processes of 'government agencies'. At the same time, it is also implying the rationality of governing citizens through their freedom. This framing chimes closely with the Foucauldian definition of the state as 'the mobile effect of a regime of multiple governmentalities' with 'the function of producing, breathing life into, and increasing freedom, of introducing additional freedom through additional control' (Foucault, 2008 [1979]: 67–8). That is not to say, however, that the possibility of coercion and the exercise of sovereign modes of power are absent from such a regime of governing. For as we discussed in Chapter 2 and as we shall see again later on, illiberal practices and sovereign impulses are still alive and kicking within contemporary biopolitical forms of governmentality. Or as Foucault himself argues, 'devices intended to produce freedom [...] risk producing exactly the opposite [as well]' (ibid.: 69). But let us, for the moment, dwell on the idea of biometric technology and identity systems as devices of control that are intended to introduce and produce (additional) freedom.

One of the most prominent forms of freedom that are increasingly targeted through such technological strategies is freedom *qua* mobility (see Cresswell, 2006; Cresswell and Merriman, 2008. In the neoliberal order of governing, mobility is considered as a vital quality and a necessary condition for promoting, actualising and optimising the ideal of active and 'flexible citizenship' (Ong, 2006: 501) insofar as it allows individuals to 'respond fluidly and opportunistically to dynamic borderless market conditions'[3] (ibid.), and enables them to enact some aspects of the neoliberal ethos of autonomy and choice. One telling example is that of the UK Iris Recognition Immigration System (IRIS). In June 2005, this biometric system was implemented in Terminals 2 and 4 at Heathrow airport, and eight months later, a similar programme was introduced in Terminal 1. The aim of this system, as stated by the Home Office, is to enable 'businessmen' and 'frequent travellers' to

> enter the UK without queuing to see an immigration officer at passport control. Instead individuals signed up to the scheme will be able to walk up to an automated barrier, simply look into a camera and if the system recognises them enter the UK, leaving immigration officers to concentrate on other priorities.
>
> (Home Office, 2006c)

Through this example, we encounter one aspect of the biometric citizen; a citizen who is capable of individual self-governing and whose freedom is expressed as freedom from 'state guidance' as well as freedom to make 'self-maximising' choices (Ong, 2006: 501) by using available advanced border technology. 'Businessmen' and 'frequent travellers' are indeed representative figures of the neoliberal subject. They are part of what became known as the 'kinetic elite', a category of mobile actors who are endowed with private[4] mobility rights and expedited border-crossing entitlements, which exempt them from waiting in busy check-in queues or undergoing lengthy security procedures. This category, as Sparke (2006: 160) puts it, 'could follow up the equivalent of a red carpet up to the border and proceed onwards with almost as little trouble as crossing a line between two provinces'.[5] Worth remembering, however, that such a form of freedom is a *conditional* freedom, one in which the privileged[6] entitlements to flexible mobility can only be obtained after submitting one's biological data and fulfilling various pre-clearance criteria that are used to assess the applicant's risk level and so on. For freedom, as van Munster (2005b: 5) puts it, 'is not just something to be let loose [...] it is also something to be managed through the constant monitoring of the things that are identified as a threat to the autonomous exercise of freedom and mobility'. As such, governing through freedom via the IRIS scheme is very much about creating the means by which freedom of mobility can be enabled, smoothened and facilitated for the qualified kinetic elite, all the while allowing the allocation of more time and effort for additional security checks to be exercised on those who are not registered on the scheme, and even more checks on those who are considered as 'high risk' travellers. In so doing, this biometric system activates a dual mechanism of categorisation and sorting; at the *virtual* level, passengers' data are pre-sorted according to those enrolled on the scheme and those who are not, and at the *actual* level, passengers' mobility is filtered on the spot within airport terminals.[7]

Therefore, in the context of freedom *qua* mobility, biometric citizenship takes on the form of what Bhandar (2004: 269) refers to as a 'privatized class of citizenship' that guarantees a privileged right of access and an 'engineered immunity from delay at the border' (Sparke, 2006: 167). The attainment and performance of this form of citizenship remain dependent upon the enterprising capacities of self-managing subjects who, as in the case of the IRIS system, *willingly* render themselves as 'flexible bodies' (Martin in ibid.) in order to achieve the benefits of this privatised flexible citizenship. Importantly, and from the paradigmatic perspective of mobility and border control through the IRIS scheme, biometric citizens are imagined not merely as holders of national passports, but more so as mobile subjects who are enacting a 'new kind of transnational para-citizenship' (Sparke, 2006: 167), a thin citizenship that goes beyond the confines of territoriality and bounded nationality. This is vividly illustrated through the enrolment criteria relating to the IRIS scheme whereby emphasis is not placed solely upon fulfilling the requirement of being a 'British citizen' *tout court*. Instead, the scheme is open to and subsumes other eligible categories including frequent short-term visitors, persons granted the right of abode in the United Kingdom on either a provisional or permanent basis, family members of nationals of EEA States and so on, provided they are able to pass the pre-clearance security and individual risk assessment checks, and other related conditions pertaining to this scheme. The scheme, however, preserves an element of contingency insofar as individuals who are enrolled on the IRIS system may still be examined by an immigration officer under paragraph 2 of Schedule 2 of the *Immigration Act 1971*, and the Secretary of State retains the right to terminate any person's participation in the scheme (see Home Office's *IRIS Scheme Definition Document*, 2002b).

Overall, what can be deduced from the case of the IRIS scheme is the fact that the practice of citizenship at the border is increasingly being rearticulated and reconfigured in terms of technical procedures of (self)government, which go beyond the traditional demarcations of nation-states and blur the distinctions entailed within territorialised entitlements (without eradicating them, nonetheless). Biometric citizenship is as such a *neoliberal citizenship* to the extent that it embodies individuated claims and practices based upon the principles of choice, autonomy, flexibility and entrepreneurialism. Ostensibly, in such a form of citizenship 'there is an act of conditioning that leads to accepting this technology [biometrics], relying on this technology, and ultimately normalizing the use of this technology' (Bhandar, 2004: 269). Control, in this context, is less about the coercive exercise of power and more about the seductive *promise* of additional freedom, privileged rights and flexible mobility. It is control in the name of freedom itself.

Such neoliberal rationality is also present in the recent identity cards scheme. The former UK government's attempt to introduce the scheme to the entire population has often been presented as a means of 'extending' and somewhat 'democratising' the reach of this additional freedom by making the convenience of speedy mobility available to the entire population, all with the underlying aim of sustaining the business-driven rationality of advanced capitalism and governing the population at distance. As indicated in the title of one of the identity speeches by Liam Byrne (former Minister of State for Immigration, Citizenship and Nationality), biometric ID cards were regarded and promoted by the former government as 'a 21st century public good' through which citizens 'from all walks of life' can manage their everyday activities and secure the integrity of their identities. Moreover, in Byrne's (2007) speech, there is a clear and strong emphasis on the rhetoric of democratisation regarding access and mobility through biometric technology. The following passage is a case in point:

> Revolutions in globalisation and technology [...] have always brought radical new possibilities [...] Modern Government's task is not to run away from that change, or shrug our shoulders in indifference, or deny its existence – but to grasp it and use it to expand horizons not for the elite, but for ordinary working families [...] But if the National Identity Scheme is to be the public good it could be, it must be accessible. The great risk of laissez-faire identity systems is the risk that they could exclude people deliberately – or price them out of secure access to things.

As such and in addition to the issue of security to which we devoted some of our earlier discussions, notions of equality and inclusion are also some of the important concepts that have been invoked and mobilised to justify the governmental rationality and motivation behind the scheme of biometric identity cards. The scheme, as such, was increasingly framed as a matter of 'inevitability' rather than choice, in the name of equalising access and generalising the convenience of technology. This form of rationalisation, as Bhandar (2004: 273) argues, 'is embedded in the compulsion that we, as the "human actors", must adjust to the demands of our biotechno-scientific capitalist society'. It represents a kind of reversed teleological order whereby the definition of the problem space itself is now often framed in terms of available technological solutions so much so that technology becomes increasingly constructed as that which *precedes* human needs rather than merely responding to them. As the following statement by Tony Blair (2006) indicates, '[t]he case for ID cards is a case not about liberty but about the modern world [....] What I do believe strongly is that we can't ignore the advances in biometric technology in a world in which protection and proof of identity are more important than ever'. So, just as one is *obliged* to regard oneself as free in the neoliberal order so as to conduct one's life according to the principles of choice and autonomy, one will now become obliged to use biometric technology with the view to 'keep up' with technological advances and make this exercise of freedom even easier, smoother and more manageable. Such modes of framing are undoubtedly behind Byrne's (2007) conviction that 'the National Identity Scheme will be just a *normal* part of British life – another great British institution without which modern life, whatever it looks like in 2020, would be quite unthinkable' (my italics).

Playing the freedom card is, therefore, a crucial aspect of the normalisation of the use of biometric identity systems and the reconfiguration of citizenship in terms of technology. And freedom, in this context, is not so much about the 'ontological' state or the 'abstract' idea of being free. It is rather something that one *does* as a matter of technical, relational and performative practices involving myriad modes of subjectification that are themselves part of a larger politics of life (see Rose, 1999: 94). In so being, freedom manages to provide at once a vehicle for legitimising mechanisms of control and a ground upon which neoliberal citizens may lay claim to their individuated rights to movement, access, consumption and entrepreneurial lifestyles. It therefore serves, through this double function, as a way of *linking* between governments, citizens and institutions, and, of course, between citizens and themselves. At the same time, the wholesale redefinition of freedom (along with associated ideals such as democracy, equality and rights) in terms of neoliberal rationalisations can also be regarded as partaking of the *disjoining* and the *disarticulation* that have taken place at the level of the concept and practice of citizenship. As Ong (2006: 499) argues, '[t]he different elements of citizenship (rights, entitlement, etc.), once assumed to go together, are becoming disarticulated from one another, and re-articulated [...] through situated mobilizations and claims in milieus of globalized contingency'. This is not only in terms of the making up of abstract 'dividuals' in the Deleuzian sense, but more so in terms of creating forms of citizenship and 'embodied' individualities that are continuously being undone and redone in their constant being-on-the-move: business travellers are re-articulating and actualising their citizenship rights according to neoliberal criteria. Asylum seekers are claiming citizenship on the ground of human rights. Skilled migrants are rearticulating their citizenship rights according to their individual performance within the workforce market. And so on. Citizenship as such is becoming like of form of collage made out of displaceable, replaceable, disjoinable and reassembleable parts.[8]

Thus, membership in a nation-state is no longer the only binding force or the major foundation for claiming political rights and entitlements. Instead, new connections and combinations are constantly emerging to respond to the fluidity and rapid changeability of the mobile globalised world, and in ways that go beyond the spatial confines of the nation state. To this end, biotechnological advancements, including those of biometrics, play a crucial role in such processes insofar as they provide viable loci for anchoring and rearticulating the myriad components that make up the different forms of neoliberal citizenship, and enable their free, mobile and entrepreneurial performance.

It is important to remember at this point that, beneath the elegant façade of neoliberal forms of citizenship and the shining veneers of entrepreneurial individualities, there lurks an important ethico-political question: 'who pays the cost of freedom for the mobility of others?' (Salter in Sparke, 2006: 169). This question immediately brings us back to some of our earlier discussions with regard to the exceptionalist power of the 'ban' that remains very much intertwined with the logic of neoliberal citizenship and with the rationality of governing through freedom. As van Munster (2005b: 5) argues vis-à-vis the EU context in general,

> the exercise of freedom within the EU does not just depend upon facilitating measures that seek to establish the optimal conditions

under which individuals actively start governing their own conduct of freedom; it increasingly also depends on the governance of what is considered to be improper and irresponsible exercise of freedom.[9]

And very often, what is regarded as improper and irresponsible, from the governmental perspective, is precisely that which stems from the movement and activities of asylum seekers and undocumented/unwanted migrants insofar as they are perceived as a threat, or at least, a hindrance to the flow of neoliberal operations. Exclusion, detention and expulsion are all some of the exceptionalist mechanisms that are intrinsic (rather than external) to contemporary modalities of governing and constitutive elements to the maintenance of their norm. For they represent a means by which residual otherness and its perceived dangerousness are regulated and contained with the aim to facilitate the exercise of freedom for those who qualify as belonging citizens and to minimise the supposed disturbance and threat of those who do not.

Unsurprisingly, the majority of policies orbiting around movement, borders and other related problem fields, are increasingly being imbued with manifold paradoxical sets of practices and phenomena. They embody an array of contradictory juxtapositions: the juxtaposition of the autonomous citizen and its risky other, of the kinetic elite and the deportation class (Salter, 2008), of flexibility and immobility, of facilitation and restriction. Increasingly, these juxtapositions function as the foundational structure and the internal logic of their strategies and rhetoric, all in the name of *balancing* and *harmonising* freedom and security for business purposes. As stated throughout the recent governmental reviews and Acts relating to borders and immigration management (Immigration, Asylum and Nationality Act 2006; Security in a Global Hub, 2008; UK Border Act 2007), there is a strong and ongoing emphasis on the need to facilitate the flow of people and goods in order to enhance the UK's image as an attractive global and economic hub, while at the same time resorting to risk management techniques and, at times, police-like forces to deter and expel the risky and the unwanted. For instance:

The UK needs strong and effective border controls to combat existing and new threats. However, this should not unnecessarily increase travel times for legitimate travellers and goods, and the disruption caused by intercepting those that represent a higher risk should be minimised. The goal is to find the optimal relationship between an appropriate degree of security, and the free flow of people and goods. Measures that discourage, or slow, movements of people and goods risk limiting the opportunities presented by trade and travel and will therefore incur a cost. However, the two objectives of security and prosperity in a global hub are not necessarily in conflict. There is significant potential for general wins through improved targeting (through better use of better information), which can lead to minimising contact and burdens on the legitimate traveller or trader while focusing impact on the illegitimate. For example, new systems that use new technology may provide a means for border control agencies to identify and fast track lower risk travellers, as well as to detain and deter higher risk passengers or goods.

(Home Office, 2008c: 28)

And again:

Through a combination of operational experience, specific intelligence and historical analysis, the Police build up pictures of suspect passengers or patterns of travel behaviour. These pictures and patterns typically share common indicators which are developed into profiles. Access to comprehensive passenger and crew data in advance of a vessel's arrival or departure in the United Kingdom will allow officers to assess the risk presented by the people carried and to mount a proportionate response. Where this involves stopping or monitoring a person or goods through the port the use of advance passenger data combined with existing intelligence systems will inform a targeted intervention, with improved likelihood of a positive outcome. A more targeted approach will also reduce the likelihood of innocent travellers being stopped, incorrect intelligence reports being entered onto Police systems and will release police resources to intelligence led activity.

(Home Office, 2006d: 37)

Sustaining a business friendly approach towards mobility and border management has thus become a matter of technically creating a successful symbiosis between freedom and security rather than regarding the two as mutually exclusive or necessarily clashing. This symbiosis itself is increasingly perceived as being dependent on creating divisions within the moving population body and producing various codified categories and hierarchisations that are amenable to different treatment and uneven regulation – the most obvious one being the division between the 'responsible citizen' and the 'abject other' (both of which subsume further divisions and fragmentations). The primary aim, as stated in the UK Green Paper *The Path to Citizenship*, is to 'widen the gap' (Home Office, 2008e: 45) between the experience of 'legal' travellers/citizens and 'illegal' migrants by making it easier for the former category to move and 'contribute' to the economy, while making it much harder for the latter to cross the border and remain in the country. In this sense, the 'logic of enabling' (movement, business, etc. for those deemed as low-risk travellers) works in tandem with the 'logic of abjection' (of those considered as risky groups) reaffirming the ongoing overlap between liberal and illiberal practices, between governmentality and exception, which underpins the overall neoliberalisation of movement and the neoliberalisation of citizenship in general. According to van Munster (2005b: 6), 'the advanced liberal notion of the free, autonomous individual has come to depend upon the abjection, exclusion and control of groups who cannot be entrusted to enjoy these freedoms'. And abjection, as Rose (1999: 253) reminds us, is primarily 'an act of force', a process of casting off or casting down persons and collectivities from a mode of existence and into a zone of shame, debasement and wretchedness. Although the force of abjection may not always be 'violence' as such (ibid.), in the case of border control, abjection takes many violent forms. In fact, and as we have seen in Chapter 2, violence at the border or, more accurately, violence *of* the border is an inextricable and inherent element of the ideal and configuring schema of the biopolitics of immigration. It is a violence that takes place at both the abstract symbolic level as well as the material embodied level.

At its most symbolic level, the violence of borders is enacted, for instance, in the act of naming; in turning singular beings into codified categories such as the 'illegal immigrant', 'asylum seeker', 'refugee', 'bogus', 'detainee', 'deportee' and the like. According to Butler (1993: 8) '[t]he naming is at once the setting of a boundary, and also the repeated inculcation of a norm'. It is, therefore, intrinsically violent in terms of both its exclusionary power and its normative character. The symbolic violence of borders is also enacted through what Balibar (2002: 143) calls the 'ultra-objective' cruelty: the 'cruelty without a face'. It is manifested, for instance, in the emergence of the phenomenon of *l'homme jetable*, the 'disposable human being' we mentioned before – a phenomenon that is increasingly being regarded, in the political imaginary, as an inevitable and thus acceptable residue of 'the establishment of a so-called New World Order' (ibid.: 142). This abstract character of violence makes violence no less violent than physical violence insofar as it instils an epistemic impulse[10] to expose singularity to 'subjection and to normalize the body politic according to the prevailing norms' (Athanasiou, 2003: 148), in ways that only foster the survival and improvement of 'the bodies that already dominate' (Diprose, 2002: 171). Biometric systems, such as the Applicant Registration Cards, are also implicated in such a symbolic violence. As discussed in an earlier chapter, not only do these systems function as a means of identification and identity verification, but also contribute to the establishment and the fixing of identity itself. No only do they institute and manage the conditions of access to social services *as* an asylum seeker, but also single out the latter as an alien, a non-citizen. Biometric systems, in this case, become a violent vehicle of interpellation through which a certain nameable identity emerges, at times, against the will of the person. Violence, as Levinas (1969: 21) argues, 'does not consist so much in injuring and annihilating persons as in interrupting their continuity, making them play roles in which they no longer recognize themselves' – such as 'becoming abject'.[11]

At its concrete embodied level, the violence of borders is unleashed in the tangible 'real' (Balibar, 2002: 141). It is the 'ultra-subjective' form of violence that manifests itself in practices of detention, expulsion and rejection. It is the violence of exposing some border-crossers to life-threatening experiences and to the labyrinth of people-trafficking, exploitation and so on. It is the violence of rendering the body as a palimpsest upon witch oppressive and unequal policies are being written, passed and endorsed, and turning the border itself into an 'exemplary theatre for staging the spectacle of the "illegal alien" that the law produces' (De Genova in Amoore, 2006: 34). Equally relevant is also the self-inflicted violence manifested through facial practices such as eyelids and lip-sewing, and mutilation of fingers (to make fingerprints illegible),[12] practices by which some 'asylum seekers' and 'detainees' assert their agency and make their ethical call heard.[13] Rather than being considered as acts of despair and distress, such practices are often portrayed in neoliberal discourses as acts of 'barbarism' and 'emotional blackmailing' that risk stirring-up the anger of tax-payers. They are regarded as proof of the inability of 'asylum seekers' and 'illegal immigrants' to 'behave themselves as independent and autonomous subjects' (van Munster, 2005b: 6). In this way, asylum seekers and illegal immigrants are also being *responsibilised* in the neoliberal order, through their presumed 'irresponsibility'[14] vis-à-vis the conduct of freedom and the ensuing failure to act as 'moral' and 'rational' subjects. 'Processes of abjection thus stress the responsibility of the abjected' (ibid.) rather than annul it, in order to reduce, discursively and materially, the responsibility of governments towards the ramifications of their policies on 'unwanted others'. Within such processes, '[a]n image takes shape – often racialized and biologized – of a permanent underclass of risky persons who exist outside the normal circuits of civility and control and will therefore require permanent and authoritarian management in the name of securing a community[15] against risks to its contentment and its pursuit of self-actualization' (Rose, 2000b: 164).

Inevitably, then, practices of citizenship remain inextricably intertwined with corrosive forms of exclusion and violence. And while violence is not necessarily the intention or the objective of neoliberal styles of citizenship, it is nevertheless a by-product, if not even a constitutive element, of the dual modality of *governing through freedom* and *governing through mistrust* inasmuch as this modality functions by means of creating caesuras within life itself and exposing the body to various forms of biopower and targeted control, some of which are deadly in their consequences (symbolically at least if not also concretely). It remains however that the key and overriding aim of neoliberal citizenship is the sustaining of market-driven ethos rather than the bringing of death into play.

Biometric citizenship as biological citizenship

Much of what has been discussed so far, in relation to the consideration of biometric citizenship *as* a neoliberal citizenship, can also be seen as unfolding in and crisscrossing with the domain of 'biological citizenship', an umbrella term covering 'all those citizenship projects that have linked their conceptions of citizens to beliefs about the biological existence of human beings' (Rose and Novas, 2002: 2). Rose's and Novas' analysis of biological citizenship is primarily concerned with the range of practices currently emanating from the life sciences and other related fields. It is an attempt to elucidate how these practices and domains are challenging traditional notions of national citizenship and thereby contributing to the development of a 'new' kind of citizenship, one in which biology plays a central role. Not that pre-existing conceptions of citizenship were devoid of biological beliefs and understandings (for biology has always been at issue in the working of citizenship projects). But what gives contemporary biological citizenship its touch of novelty, according to Rose and Novas, is the fact that it does not necessarily take a 'racialized and nationalised form' in order to reach racial purity – as was the case during the eugenic age for instance. Instead, biological citizenship projects take as their task the maximisation of what Waldby refers to as biovalue, that is to say, the rendering of life itself as a productive and profitable economic and political value.[16] While Rose and Novas are not directly addressing the issue of biometrics when invoking the notions of biovalue and biological citizenship, I use these concepts to capture and illustrate some of the dynamics at work within the rationalities and practices of biometric technology. By the same token, and contrary to Rose's and Novas' assertions, I demonstrate how the racial and nationalised aspects are not entirely absent from biological citizenship but take on different, and at times implicit, functions and dimensions.

As discussed previously, the deployment of biometrics within the fields of border and immigration management functions as a biopolitical technique for sifting through different forms of life according to their level of usefulness and legitimacy in order to define and distinguish between those who can contribute to the economy and those who have little or nothing to contribute. It thereby provides a means of organising and categorising individual and collective biovalue by opening up the body to various economically driven processes of sorting and turning it into an anchoring point of reference for linking the person's identity to her biovalue (as with the case of work permits and skilled migrant programmes). In the recent governmental proposals, concerning the reform of the current citizenship system in the UK, the use of biometric technology and identity systems feature quite prominently. In *The Path to Citizenship*, for instance, the Home Office (2008e: 47), advocates the legal provision of the following:

- One comprehensive power to obtain and use biometric information in the situations and from the classes of individual where there is a need to do so.
- This power would allow the [Border and Immigration] Agency to obtain biometrics to verify identity, or if the individual is not already known to us, to establish identity, in potentially every situation in which we come into contact with an individual.
- Gateways to share this information with other bodies for other purposes.

Citizenship, as such, is increasingly being formulated in technological rather than territorial terms, as that which relies, for its management, upon the enactment of the power of obtaining, storing, sorting and sharing biometric information. Whether as an ideal or as a practice, citizenship is being reduced, as Muller (2004; 2010) argues, to processes of identity management whereby the convergence of body and technology function as an important means for assembling, distributing, managing and actualising emerging political claims and citizenship practices. Moreover, citizenship is also increasingly being regarded as a kind of 'deal' (Jacqui Smith in Home Office, 2008e: 5) insofar as '[m]igrants must "pay their way" in order to qualify to be a citizen/permanent resident' (Home Office, 2008e: 4). Such formulations confirm the government's devotion to neoliberalising and biologising citizenship all the more by making 'rights and benefits contingent upon individual market performance' (Ong, 2006: 500) and by fixing identity in terms of biological characteristics. In this respect, one can regard biometric citizenship as being situated at the crossroad between neoliberal citizenship and biological citizenship inasmuch as it knits together biology and capitalism while locating identity in the biopolitical sphere of governing.

Correlatively, biometric citizenship also marks a shift towards the neoliberalisation and biologisation of the concept of *homo-œconomicus*, that is, man as an economic subject. For Foucault (2008 [1979]: 225–6), this figure has witnessed an important mutation since its classical conception insofar as the *homo-œconomicus* is no longer merely perceived as a 'partner of exchange' who partakes of the problematic equation of needs and utility. Instead, within the neoliberal rationalisation, *homo-œconomicus* has become a 'man of enterprise and production', 'the vis-à-vis, and the basic element of the new governmental reason' (ibid.: 147, 271), a man who strives to produce and maximise his own satisfaction through various means and technologies of the self. *Homo-œconomicus* is the neoliberal subject *par excellence*, the citizen-consumer who performs according to market principles and is constantly engaged in endless transactions to guarantee maximum benefits and secure self-interest.

In a way, the voluntary use of biometrics for the purpose of economic and mobility facilitation and optimisation, as in the case of the IRIS system, can be regarded as an instantiation of some of the technologies of the self by which *homo-œconomicus* seeks to exercise the freedom of mobility, choice, lifestyle and so on. And, by virtue of being 'eminently governable' (ibid.: 270), *homo-œconomicus* is not hesitant to put his body at the service of technology if that opens up opportunities and ensures the criteria of efficacy for actualising and facilitating the performance of such freedoms. At the same time and in terms of the government's relation to *homo-œconomicus*, biometrics can be seen as a way of biologising the citizen *qua* economic actor and thereby reinforcing the link between the neoliberal and biological aspects of citizenship. The caesuras that are introduced into the life of the (moving) population though techniques of biometric control are, in this sense, played out at the very ontological level of what it means to be human. According to Bhandar (2004: 270–1), 'the shift to *homo-œconomicus* refers also to a shift in the foundational status of what is *human*. The category of human has never been a historically universal or all-inclusive category, but rather has operated through systematic technologies of inclusion/exclusion'. Whereas in the traditional models of citizenship, the spatial partitioning together with its national dimension have been the primary means of demarcating between the included and excluded, in the neoliberal order, it is the bio-economic factor that tends to separate between these categories (though the former remains in place as well).

It should be borne in mind that the biologisation of the (non-)citizen for economic purposes does not entirely purge contemporary formulations and practices of citizenship from the racialised and nationalised dimension, nor is 'economic triage' the only form of selection that drives immigration policy. At one level, biological markers are now being put to the service of different forms of racialisation, often leading to what Balibar (1991: 21) calls 'racism without races'. This, to the extent that the category of immigration is increasingly functioning as 'a substitute for the notion of race' (ibid.: 20) and the ground upon which various modes of discrimination and xenophobic activities are routinely exercised in the name of security and counter-terrorism policies. In recent years, there has been an increase in 'e-borders' schemes worldwide, whose primary function is the gathering, analysis and sharing of information relating to passengers prior to their travel. As stated by the Home Office (2008c: 48),

> The work of the border agencies begins long before a person or consignment arrives at the physical UK border. Visas are issued (or refused) overseas, some specific HM Revenue and Customs (HMRC) checks are undertaken, and data relating to passengers and goods due to travel to the UK are analysed to determine whether they should be subject to intervention on arrival or even be allowed to undertake travel at all [...] Some individuals are known to represent a risk. Others seek to travel under identities that are known to be unreliable. Early action is possible in both situations. The infrastructure needed to address this risk is shared: a single pool of information about suspect identities and risky individuals; fast and secure access to that information by authorised officers; and comprehensive coverage of

those individuals travelling to the UK [...] The UK already collects data on those travelling from countries judged to present the highest risk through visa regimes. This is supplemented by, for instance, advance passenger information and reservation data on routes via Project Semaphore.

With such pre-emptive techniques, racial, ethnic and religious profiling of those considered as 'risky' individuals and collectivities becomes an almost inevitable outcome of the merging of technology, security and what Delsol (2008: 2) refers to as 'the habitual and subconscious use of negative stereotypes – stereotypes that are deeply rooted in the institutional culture of law enforcement' (see also Côté-Boucher, 2008). In the US case, for instance, the Computer Assisted Passenger Pre-Screening system is used to collect data about all passengers travelling to the United States and classify them into three different categories/identities according to the level of their potential risk, parameters of which include the element of race (Lyon in van Munster, 2005a: 10). In Europe, the 'stop and search' activities against illegal immigrants are often conducted on the basis of 'appearance' targeting minorities and singling out specific groups for further investigation. Race, as such, is increasingly functioning as a proxy for nationality (Delsol, 2008: 6) the same time that the category of immigration is increasingly becoming the surrogate signifier of race. As stated by a senior Spanish officer, '[w]e stop foreigners to see if they are illegal; how can we enforce the [immigration] law if we don't stop people who look like foreigners?' (in ibid.: 5). These explicit practices of racial profiling are part of the banoptic dispositif in which '[a] skin colour, an accent, an attitude and one is slotted, extracted from the unmarked masses and, if necessary, evacuated' (Bigo, 2006b: 44).[17]

In May 2005 and in response to the growing concerns over the potential implications of the identity cards scheme vis-à-vis the issue of racial profiling and discrimination, the Home Office published the *Race Equality Impact Assessment*. The report argues that the scheme itself, given its 'inclusive' aspect, is non-discriminatory in that it is intended to cover *everyone* who is residing in the United Kingdom legally for longer than three months. Since then, the delivery plan of the scheme has undergone myriad changes and updates. In 2008, powers from the *UK Borders Act 2007* have been used to introduce compulsory identity cards for foreign nationals who are subject to immigration control. The phased roll-out strategy of the delivery of ID cards started with the enrolment of foreign students and those on marriage visa, since they are considered by the government as constituting 'high risk categories' in terms of the abuse of the immigration system (see Home Office, 2008d). The roll-out strategy has later on extended to other 'high-risk' immigration categories including Work Permits holders and those applying as a child of a foreign national parent already settled in the UK (ibid.). The aim is to cover all foreign nationals by 2014/15. In February 2011 and under the Coalition Government, the UK national identity card scheme has been abolished. ID cards are now compulsory for foreign nationals only.

Changes of this kind make the argument about the non-discriminatory inclusiveness of the scheme rather redundant. They in fact indicate a direct and explicit discriminatory character insofar as the selection of specific groups for compulsory identity requirements and their designation as high-risk categories open the door to various prejudicial and differential treatments. This in turn can have many negative ramifications touching 'the lives of the weakest and most marginalized members of the population' (Lyon, 2007: 115). In response to the government's *Compulsory Identity Cards for Foreign Nationals*, the Northern Ireland Human Rights Commission provides a cogent assessment of the impact of the scheme with regard to the issue of race and profiling. One of the main and recurring concerns expressed by the Commission is to do with the nature of the discourse used by the government to promote the ID cards scheme. This discourse, according to the Commission, contributes to the stigmatisation and demonisation of foreign nationals, and to the negative media reporting around the issue of immigration and the inciting of public fears:

> Care should be taken that in compliance with this recommendation and other obligations the government's own discourse in promoting the scheme does not contribute to demonization of migrants. Language such as that contained in the Ministerial Foreword to the present consultation, which includes a metaphor comparing tackling irregular migrants to dealing with effluent ['flush out those who evade our rules and laws'], is clearly inappropriate.
>
> (Northern Ireland Human Rights Commission, 2008a: 4)

[i]n assessing the UK's compliance with its human rights commitments the Council of Europe recently raised concerns that 'Negative and inaccurate reporting by sections of the media is contributing to hostile attitudes towards certain groups'. It singled out discussion on groups including asylum seekers, migrant workers and Muslims, arguing that sections of the media often discuss such groups 'in a

manner that is often biased, stereotyped and inaccurate' and raised concerns that this contributes 'to a climate of fear and hostility and aggravating community relations'.

(ibid.)

Moreover, the lack of clarity with regard to the use of the term 'compulsory' makes the latter amenable to various interpretations, one of which might be translated into 'a "papers please" environment whereby foreign nationals compelled to register for the card are expected to carry it with them at all times' (ibid.: 6). This in turn may lead to the development of 'a culture of suspicion against foreign nationals (or perceived foreign nationals) who do not permanently carry the card on them, even when there is no requirement to do so' (ibid.). With this, the Commission argues, it is likely that individuals who 'do no look like' EEA nationals will be asked to produce identity cards more frequently, raising the disproportionate numbers of stops and searches conducted against individuals belonging to ethnic minorities and creating an 'effective police state' for migrants. The Commission also expresses its concern regarding the subjection of children to the foreign nationals' compulsory identity scheme insofar as this 'may amount to punishing children for the parent or guardians' immigration status' (ibid.: 2), and thereby undermining some of the government's obligations vis-à-vis the UN Convention on the Rights of the Child.

From the above, it is clear that racial discrimination is a risk that looms heavily over the scheme of compulsory identity cards for foreign nationals. This risk is not only concerning the ID card itself but the entire informational system of the scheme wherein 'the real surveillance power lies, to discriminate between different categories and groups, for differential treatment' (Lyon, 2007: 112). For without the necessary safeguards, similar profiling trends that have been taking place through European immigration databases, such as those of Eurodac, Visa Information System and the Schengen Information System, may be performed in the UK as well. As regards biometric techniques, their special trick, according to Gilbert (2007: 90), is that 'they obfuscate the very embodied dimensions of their classification by turning instead upon languages of authenticity and inauthenticity (see Muller, 2004). So while the body is used to fix an identity, the racialization and biologization of these discourses is obscured'.

As discussed in Chapter 1, biometrics is not without a history. It is embedded within a colonial past whereby race has functioned as a prominent component within the mechanisms of power and control (see also Maguire, 2009; Muller, 2010; Pugliese, 2010). Race remains, as Stuart Hall (1993: 298) argues, as an 'organising category of [...] ways of speaking, systems of representation, and social practices (discourses) which utilize a loose, often unspecified set of differences in physical characteristics [...] as *symbolic markers* in order to differentiate one socially from another'. With contemporary remediations of biometric techniques, and despite the technical camouflaging of their racialised and biologised aspects, there is a strong sense in which the very technological infrastructure of biometrics is inscribed within a race-centred discourse and practice to the extent that it is calibrated on 'whiteness' as a universal category. This point is taken up skilfully by Pugliese (2007) in his consideration of the intersection between biometrics, bodies and race. Central to his thesis is the argument that 'Western technologies of representation have a long history of setting the operating infrastructure of these technologies according to white templates' (Pugliese, 2007: 107). This is evident in the case of biometric technology whereby whiteness is set as 'the universal gauge' that establishes the operating parameters and determines the technical measures for the visual capture and imaging of the subject (ibid.). In this sense, whiteness is seldom perceived as a racial category in itself. For it is so diffuse within the fabric of everyday life and inscribed within the fibre of its myriad technologies. As eloquently put by Pugliese (2007: 107),

> One cannot talk of whiteness as such, as the power of whiteness resides in this capacity to occlude and so mystify its status as a racial category that it too often escapes taxonomic determination, while simultaneously remaining the superordinate racial category that effectively determines the distribution of all other classificatory categories along the racial scale.

The concrete manifestation of this argument can be witnessed through the instances whereby biometric technology 'fails' to capture and enrol certain bodies precisely because of the subject's race, which does not conform to the encoded white standards of biometric operations. For example, biometric fingerprint scanners have routinely encountered difficulties in reliably capturing and verifying the fingerprints of Asian women because of their 'fine skin' and 'faint' fingerprint ridges, while dark-skinned users are not easily 'distinguished' by facial-scanners (ibid.: 112).[18] In Japan, a study has argued that biometrics would find it most difficult to identify 'non-Japanese' faces (Tanaka et al. in Magnet, 2007). These functional failures are indeed epitomes of the way in which biometrics industry is infrastructurally embedded in ethnic as well as demographic stereotypes, rehearsing some of the racialised and colonial discourses regarding the 'inscrutability' of certain bodies. As the following statement form the scientific literature indicates:

> Facial-scan systems' sensitivity to lighting and gain can actually result in reduced ability to acquire faces from individuals of certain races and ethnicities. Select Hispanic, black, and Asian individuals can be more difficult to enroll and verify in some facial-scan systems because acquisition devices are not always optimized to acquire darker faces. At times, an individual may stand in front of a facial-scan system and simply not be found. While the issue of failure-to-enroll is present in all biometric systems, many are surprised that facial-scan systems occasionally encounter faces they cannot enrol.
>
> (Nanavati et al. in Pugliese, 2007: 113)

Therefore, it seems that within this economy of visual representation, there is a deep-seated oblivion and a lack of reflexivity regarding the ramifications of the technology–race nexus.[19] Instead of taking into account the causal relationship between these functional failures and the techno-ideological encoding of whiteness as the universal norm, some bodies are simply portrayed as being *invisible*, *irregular* and thus *problematic* bodies that cannot be 'found' by the system and whose difference is incomprehensible to biometric procedures due to their chromatic or featural deviation from the normative zone of whiteness. It is as though these non-white subjects are ghosts in the biometric machine whose presence is an absent presence and whose appearance is a nonappearing appearance. Such occlusion, according to Pugliese (2007: 113), 'gestures toward a racialized zero degree of nonrepresentation' while creating an onto-epistemological split that 'pits subaltern being against elite knowing' (Spivak in ibid.: 118). At the same time, these systemic and discursive processes of occlusion, irregularisation and invisibilisation of certain bodies are symptomatic of the mythical character of the much-vaunted race-neutrality and technical impartiality to which biometric technology lays claim.

To be sure, when it comes to the current rhetoric and practices of biometric citizenship in general, myths abound. One notable myth is to do with the nationalised aspect that is also intimately linked to the racialised dimension. After all, and as Balibar (1991: 37) argues, 'the discourses of race and nation are never very far apart, if only in the form of disavowal'. In the case of ID cards, this takes the shape of a symbolic capital that has been attached to the scheme, making ID cards stand as a token of Britishness itself rather than merely as a convenient technical system of identification. Arun Kundnani (2008), form the Institute of Race Relations, picks up on this point when examining New Labour's arguments for ID cards. He states:

> When the idea of ID cards was first introduced, they were described as entitlement cards. Very quickly, an idea that ID cards could be an emblem of national identity took hold, and, particularly, look at the writings of David Goodhart (the editor of *Prospect* magazine and somebody who is very close to Home Office thinking on this), who has argued very forcefully that ID cards can be a way of giving citizenship a practical meaning, so you hold up your ID cards and say, 'I am proud to be British'.

The imposition of 'compulsory' identity cards on foreign nationals does indeed introduce a paradox at the heart of this symbolism in that it challenges the government's initial argument that, in order for identity cards to be non-discriminatory, they have to be made compulsory for everyone in the UK rather than for just specific groups:

> [i]f you say to Goodhart or to ministers that ID cards are going to introduce a massive new layer of discrimination against minorities, their response is, yes, you are right, that is exactly why we need to make them compulsory for every British citizen, so everyone can be asked for their ID cards, not just minorities.
>
> (ibid.)

So, although the abolition of the national ID cards scheme under the current government is seen as a momentous positive step towards restoring and safeguarding civil liberties, it also represents a negation of what was once the most forceful argument regarding the inclusive and non-discriminating nature of the scheme since ID cards are now compulsory for foreign nationals only.

Furthermore, the recent governmental proposals regarding the rules of progression between different immigration statuses and into full citizenship also indicate a nationalised component despite their overemphasis on the economic aspect. In *The Path to Citizenship*, for instance, the government makes repeated references to and a strong argument for reinforcing 'British values' and placing them at the heart of the immigration system: '[w]e believe we need to work harder to strengthen the things – the values, the habits, the qualities – that we have in common, and thereby strengthen our communities' (Home Office, 2008e: 12). There are of course many inherent problems in promoting such an agenda. First, it is unclear as to what exactly a common set of 'British values' might be.[20] Apart from 'the NHS [...] our values of tolerance, fairness and freedom of speech, a healthy disrespect for authority and yet a keen sense of order' (ibid.: 14) – elements that are by no means distinctively British, the government makes no clear indication as to what it implies by the common 'values, habits and qualities'. This, in turn, leaves such a statement open to various interpretations, one of which might imply 'convergence in outlook and behaviour', which 'is likely to limit diversity, by making migration more difficult for those who do not fit with a presumed "British" way of life' (Migration and Law Network, 2008). Or, as the Northern Ireland Human Rights Commission (2008b) puts it, '[t]he tone of the proposals could be interpreted as British citizens holding a particular set of values that are not shared by non-Europeans and need to be nurtured or taught', which then opens the government to accusations of 'colonial discourse' (ibid.).

We can, therefore, see through these discussions that, despite the neoliberalisation of citizenship and its attendant disarticulation of the figure of the nation-state, the myth of a common national identity is still subsisting within the governmental rationalities and discourses surrounding the issues of immigration, citizenship, community and belonging. It is a myth that both feeds into and is fuelled by the identitarian zeal for communal sameness and whose essence carries, implicitly if not explicitly, that insidious form of 'racism without races' wherein the

> dominant theme is not biological heredity but the insurmontability of cultural differences, a racism which, at first sight, does not postulate the superiority of certain groups or peoples in relation to others but 'only' the harmfulness of abolishing frontiers, the incompatibility of life-styles and traditions; in short, it is what P.A.Taguieff has rightly called a *differentialist racism*.
>
> (Balibar, 1991: 21)

Biometric citizenship as neurotic citizenship

So far throughout this chapter, we have examined how the governance of identity and citizenship has become increasingly reliant on the 'rational' capacities of the self-actualising citizen. Governing through freedom amounts to calling upon the entrepreneurial spirit and the myriad qualities of the neoliberal subject and mobilising various biotechnological mechanisms of self-management and self-policing. At the same time, this also amounts to pre-empting the (perceived) risks of those who are considered as a threat to the exercise of freedom and to the smooth running of business operations. As such, 'the distribution of trust and fear has a fundamental impact upon how freedom is distributed' (van Munster, 2005b: 22), which makes governing through freedom inextricably linked to governing through mistrust. In what follows, I shall argue that within the nexus of these two modalities lies another form of governing, namely *governing through affects*.

Invoking the notion of affects in relation to governmentality immediately calls into question the familiar image of the rational subject. This image has been so predominant within neoliberalism and crucial to its overall materialisation. For governing, in its various forms and modulations, has long been based upon the ability to make rational and calculated decisions vis-à-vis future choices and actions. To this end and throughout the decades, the figure of the (neo)liberal citizen has been imagined within governmental strategies as that which regards being rational as a paramount and necessary ingredient for conducting one's life successfully and productively, and for handling the risks and uncertainties of the future. It is, however, the nature of the future to bring about risks and uncertainties that are not always amenable to *full* calculation and *complete* pre-emption. Some *events* belong to that radical realm of the unforeseen, the unknown and the unintended wherein one is inescapably confronted with the limits of knowledge, rationality if not even reason itself. And as techno-scientific developments, global movement, business ventures, 'terrorist' threat and military actions increase in reach and magnitude, so too does the uncertain and the unknown. Consequently, the problems of governing 'start to extend beyond the foreseeable to the unforeseeable', raising anxieties about the risks pertaining to 'the unpredictable and uncontainable impact of human actions' (Diprose et al., 2008: 269). Citizens are therefore encouraged to be prudent 'to imagine the unimaginable and to actively prepare for protecting themselves from a future threat' (ibid.: 267–9) insofar as uncertainty is considered as 'no excuse for inaction' (Stern and Wiener in de Goede and Randalls, 2009: 860).

In this prudentialist approach towards the future, the rational subject seems to be accompanied with another kind of subject: a subject that is constantly anxious and continuously caught up in a cycle of hyper-vigilance that borders on paranoia. The fact is that 'the proliferation of an everyday culture of risk places burdensome demands upon the self, forcing individuals to habitually make reflexive choices' (Mythen and Walklate, 2006: 383). The neoliberal citizen, as such, keeps oscillating between these twin poles of subjectivity whereby he is expected to act rationally in the midst of his anxieties and, at the same time, cannot seem to escape his anxieties despite his endeavour to sustain his rationality. He is left to wonder why, despite all the responsibilities he took upon himself, despite all the self-management techniques he has acquired, despite all the prices he has paid, he is still left dissatisfied, scared and restless; a thorny and unenviable situation, no doubt. For what could be more unsettling to a self-assured neoliberal citizen than being reminded of his limits and vulnerability? And so, the more he attempts to rationalise his anxiety, the more anxious and irrational the neoliberal citizen becomes. He is then faced with the task of governing yet another aspect of himself: his affects.

> 'biopolitics' does not adequately capture or account for the subject who is governed through its affects. At the centre of biopolitics was what I call the bionic citizen, a subject whose rational and calculating capacities enabled to calibrate its conduct [...] By interpreting the liberal and neoliberal subject as the bionic citizen, who was self-sufficient, self-regarding and was governed in and through its freedom, we may have unconsciously participated in the production of a phantasy.
>
> (Isin, 2004: 222)

Indeed, the sheer majority of work that has been conducted in and on the area of governmentality and biopolitics tends to hinge mainly upon the image of the rational subject. It too seems to have been caught up in this exaggerated way of framing the citizen as competent, autonomous, responsible and entrepreneurial. Recently, however, the dimension of affects and emotions have started receiving some attention within governmentality and citizenship studies, destabilising some of the earlier assumptions regarding the figure of the rational subject. Nevertheless, the calling into question of the latter through the consideration of affects amounts to more than its mere destabilisation. It is also aimed at elucidating how affects themselves are becoming prominent vehicles for governmental rationalities (ibid.: 223) and how anxiety, in particular, is emerging as 'an increasingly powerful political force' (van Munster, 2005b: 22).

In his essay 'The Neurotic Citizen', Isin (2004) provides an insightful account on the ways in which subjects are being governed through their responses to anxieties and uncertainties and encouraged to manage and calibrate their conduct on the basis of their insecurities rather than rationalities. He takes cue from the psychoanalytic thought whereby 'neurosis' is regarded as 'an inescapable condition of existence because the subject, caught up in its identification with an illusory, unattainable imago of wholeness and in its ultimately unfulfillable desire, could never attain a sufficient wholesomeness that is always posited as "normalcy"' (ibid.: 223). Isin's deployment of this discourse of neurosis is not meant to pathologise the subject of government, but serves as a mechanism to describe the inevitable and ongoing psychic struggle facing the latter. This struggle, according to Isin, stems mainly from the subject's desire to attain and possibilise the impossible – such as absolute security, absolute safety, absolute tranquillity and so on, states of being that hardly match up with the external reality and its escalating culture of fear and dominant climate of insecurity and suspicion. Isin coins the term 'neuroliberalism' to designate a mode of governing though neurosis whose subject is 'less understood as a rational, calculating and competent subject who can evaluate alternatives with relative success to avoid or eliminate risks and more as someone who is anxious, under stress and increasingly insecure' (Isin, 2004: 225). That is not to say that the rational subject and the neurotic subject are mutually exclusive or amenable to dichotomous separation. Rather, and through the intersection of the different modalities of governing, citizens embody both of these subjectivities and appear to be implicated in a double and intricate movement by which they are, on the one hand, placed under 'the pressure to appear normal' (Salter, 2008: 374) while, on the other hand, they are incited to conduct themselves as neurotic citizens (Isin, 2004: 226).

There exist a myriad of sites and practices that are illustrative of the ways in which citizens are continuously being neuroticised and imbued with a heightened level of stress and anxiety. Isin invokes a number of examples ranging from the economy and the environment to the body and the computer-based networks. To be sure, no domain seems to escape the clutches of neuroses and be immune to their accompanying affects and effects. Even the home, a site that is supposed to represent a safe haven and provide freedom from fear, insecurity and instability, is ironically turning into a space where further anxieties and insecurities are produced and managed through home surveillance technologies and the architectural structure of gated communities (ibid.: 230–1). Of particular relevance to our discussion is the example of the border. In fact, what sparked Isin's interest in the figure of the neurotic citizen in the first place was 'the anxiety about the Other that has been articulating itself through various discourses on the border and which has gathered strength and reassembled itself since the events named after a month and day' (ibid.: 231). But as Salter (2008: 374) argues,

> not only have these discourses of anxiety and neurosis been dominant since 9/11, but also these affective politics have been in play to some degree or other since the consolidation of the territorial-sovereign state. The pressure to produce a truth for the representative of the sovereign – a truth which only that representative may authorize – has been essential to the construction of borders.

Salter also identifies the 'dialectical' feature of the affective politics of borders. For while neurosis is mobilised as a prominent resource of governing at the border, there is also 'a marshalling of pleasure as a political resource in the correct presentation of authentic documents, and the telling of an acceptable story' (ibid.). The IRIS system, for instance, provides a valid example of the kind of pleasure that can be attained at the border through the sense of privilege and the convenience of expedited crossing, illustrating the luring capacity of biometric technology. Akin to a reward for obedience (ibid.), the experience of pleasure at the border is thus brought into accord with the wider dynamics of security, power and control. But let us remain with the issue of neurosis for the moment and examine its interplay with biometric citizenship at the border.

Some of the latest developments in biometric technology for border security have been focusing on 'behavioural profiling': a form of profiling that takes affective expressions as its object of analysis and scrutiny. According to Adey (2009: 275), such developments gesture towards an ironic mood shift in surveillance and security whereby attention is now being increasingly directed towards 'microscopic particles and traces', 'physiological indicators' and 'microscopic gesticulations'. Perhaps not so much a 'mood shift' per se, but a *continuation* and an enhancement of already established modes of surveillance and techniques of securitisation. For in addition to measuring the body's exterior surface (fingerprints, iris scans, hand geometry, etc.), these biometric mutations also aim to capture another dimension of the mobile body by delving into one's interiority and mapping out the internal and external interplay between intentions and feelings. They are geared towards the pre-emptive securitisation of the unconscious and the predictive theorisation of the unknown whose objective, as we have discussed before, is the monitoring of the future and the anticipation of action. In so doing, 'they construct an imagination of an anxious and neurotic subject of drives, instincts, moods, and emotions' (ibid.: 275).

As such and within the context of the spatio-temporal zone of borders and airports, emotion in motion is increasingly perceived as an important resource for security procedures and a potential target of biometric control. Not that the focus on emotional and affective aspects in the context of mobility is new in itself. But, as Adey (2009: 278) argues, airline, airport and security professionals have long been interested in pre-visualising and imagining the passengers' needs and wants in order to predict and anticipate their behaviour. 'Such an attention has arisen in part so that the consumerist political economy of the airport terminal can be managed at a profit as passengers' "felt experiences" are made both measurable and quantifiable' (ibid.). Once again, we are encountering another instance whereby economic interests are being merged with security interests within a single, but nonetheless multidimensional, paradigm of control.

It is important to point out here that, with regard to biometric behavioural profiling, the aim is less about capturing the entire spectrum of affects relating to the mass moving body and more about searching for biological and gestural indicators of instinctive and uncontrollable 'emotion leakages' such as guilt, anxiety and fear of being caught, feelings that the body cannot always contain or conceal (ibid.: 284). The concern is with pre-social reflexes and primal instincts and drives that are considered as intrinsic and common to all humans. The filtering and sorting potential of behavioural profiling lies in its ability to sift through what is considered as 'normal' legitimate fears associated with travelling (after all, airports have always been emotional spaces filled with circulating fears, stress, dread, tension, boredom, excitement and so on) and the more intense and thereby 'suspicious' anxieties that might be linked to hostile intents. The focus of this form of profiling is on what Tomkins calls 'affect-about-affect' or what we may also refer to as 'reflexive affects', 'emotions generated by the generation of their initial affective state' (Adey, 2009: 286) (e.g. being anxious about feeling anxious, being afraid of one's anger). In the barometer of the biometric behavioural profiling apparatus, these reflexive affects are assigned levels and degrees of intensities in order to detect amplified emotions and thereby determine 'elevated behaviours' that deviate from the modelled normative 'baseline behaviour' (Frank in ibid.: 285).

In this sense and within such technical imaginary of control, not only the exteriority of the body and its genetic code that are used as a 'witness against oneself' (van der Ploeg, 1999b) but also the body's internal rhythms and physio-affective transmissions and circulations. So while in biometric identification, the physical body is used as a biological marker of *identity*, in behavioural profiling, the body's automatic responses and impulses are used as indices of *intent* – although in both techniques, the body itself remains the vehicle of transmission and the link between the interior and the exterior. At work here is the institutionalisation of anxiety at the border (Salter, 2008: 376) and with it the activation of another layer of confession that is not only extracted from the singularity of body parts, but from the expressions and responses triggered by reflexive affects that are themselves auto-affective products of 'nervous pitches and tones' (Adey, 2009: 287). For example,

[t]he device, dubbed MALINTENT by inventors, uses sophisticated sensors to read body temperature, heart rate and respiration. Analysed together, these factors can lead security services to potential terrorists. Any suspects are pulled aside for questioning and then subjected to a second scan, which involve micro-facial scanning. This equipment is able to read minute muscle movements which give further indications of criminal intent. So far it can recognise seven primary emotions and emotional clues and will eventually have equipment which can analyse body movement, an eye scanner and a pheromone-reader. More importantly, developers have programmed it to recognise the difference between someone who is simply stressed and a potential terrorist.

(Hazelton, 2008)

The increasing interest in emotions and their embodied manifestation represents another symptom of the desire to intensify border security and control, a trend that some are referring to as the will to 'Israelify' borders and airports. As one journalist puts it: 'one word keeps popping out of the mouths of experts: Israelification. That is, how can we make our airports more like Israel's, which deal with far greater terror threat with far less convenience' (Kelly, 2010). And from the point of view of security industry experts, Israelifying airports and borders involves, in part, the convergence of different techniques as a way of promoting a 'holistic' approach towards security through the technical and epistemic triangulation of body, emotion and thought. For instance, SDS-VR-1000, a machine developed by the Israeli security firm Suspect Detection

Systems (SDS), is designed to record and measure the facial and physiological responses to questions in order to detect suspect behaviour and hostile intent:[21]

> What [the machine] does is collect objective data out of the passenger's ID – and it analyzes the data compared to the subjective data it collects while the passenger is asked different questions. [T]he process takes about three minutes, and the passenger either receives a transfer printout authorizing him to advance to the next stage of entry to the country, or an announcement that he is required for further questioning. A monitoring official will then escort the passenger to another area for further questioning.
>
> (Shoval, 2005)

And as Adey (2009: 287) explains further:

> The questions posed usually encompass particular words that are intended to agitate the guilty respondents and activate certain bodily responses. SDS has developed a word library which it believes only terrorists will respond to. These include words that name specialised materials relevant to terrorist activities such as the making of a bomb. Hence, 'semtex' could set off a particularly nervous reaction. Or the kinds of terminology that refer to the mental and spiritual preparation suicide bombers may undergo prior to their attack may be referred to. The machine observes changes in vocal pitch and other indicators of stress and disorientation.

Bringing 'narrative' and 'affects' (back) into the border security equation is in a way an attempt to fill the gap within the automated biometric scanning systems.[22] As I discussed in Chapter 3 and in the context of asylum management, biometric techniques that seek to capture the singularity of the body for the purpose of identification and identity verification cannot fully securitise the border against the perceived threats, nor can they reveal intentions and the like insofar as they exclude narrative and with it the affective dimensions. To this end, behavioural detection and profiling may be regarded as a complement to biometric identification as well as a continuation of what Adey (2009: 288) refers to as 'a touchy-feely form of security concerned with identifying and thereby securing the well-being – the feelings – of its population. But [it] is the feelings of the population that are used as the very means of security'. This in turn raises additional sets of questions and concerns regarding the ethical implications of governing through affects whereby narrative, as in the case of SDS-VR-1000, is not used as a means of creating an ethical space of solicitude (as I tried to propose in Chapter 3) but as a confessionary and coercive tool for extracting the truth about one's intentions and thoughts. 'Soul-scanning' in addition to 'body-scanning' is what seems to be on the security agenda, at the moment. And whether in terms of biometric identification or behavioural profiling,

> [this] will to control time and space, present and future, here and there, has an effect that goes beyond antiterrorist policies; it creates a powerful mixture of fiction and reality, of virtual and actual, which merge their boundaries and introduce fiction into reality for profiling as well as it de-realizing the violence of the state and of the clandestine organizations.
>
> (Bigo, 2006a: 62)

Another quintessential example and a more 'routine' site for understanding the interplay between affects and biometrics is that of identity and its management against the threat of fraud. In a previous section, we looked at the ways in which identity is being securitised against this threat through the deployment of various risk management techniques and technologies, including biometrics. We argued that within the discourses and strategies of precaution and pre-emption surrounding this issue, the risk is not only articulated in terms of its financial effects but also in terms its emotional ramifications. We shall now extend our previous discussion and examine how affects are alarmingly mobilised so as to encourage citizens to actively protect their identities and manage the adverse emotional side effects of falling victim to identity fraud and theft.

In October 2009, the UK's Fraud Prevention Service (CIFAS) published its report *The Anonymous Attacker*. The report combines statistical updates on the scale and growth of crimes relating to identity fraud, tips on how to prevent the occurrence of fraud and testimonies from victims about its financial and emotional effects. It begins with the following statement: 'It is a fact that fraud increases during recession' (CIFAS, 2009: 2), a statement that echoes the continuous governmental attempts to fashion a 'hyper-vigilant' subject (see also Whitson and Haggerty, 2008: 256) and place the risk of identity fraud within a seemingly up-to-date context. For, faced with this 'fact', the reader is immediately injected with a sense of prudence and made aware of the perceived correlation between identity fraud and the current economic climate: '[a]s credit granting diminished during the credit crunch, fraudsters saw the writing on the wall and turned their attention away from application fraud, knowing that this would present fewer and fewer opportunities. Instead fraudsters have reacted to economic circumstances, and migrated towards identity fraud and account takeover fraud' (ibid.: 4). After making various quantitative statements about the prevalence of identity fraud and its current patterns, the report continues with a word of caution against negligence and the lack of proactivity in dealing with what is considered as the fastest growing crime in the UK:

> Despite repeated warnings, people are still not taking precautions to safeguard their identities, which is why new figures reveal that the UK has the highest number of victims in Europe. We all have the responsibility to ensure that no confidential information, whether business or personal, leaks into hands of fraudsters [...] So prevention is definitely better than cure! [...] Individuals therefore must be vigilant, and must take responsibility for protecting their own identity to the best of their ability.
>
> (ibid.: 3–4)

Responsibilisation and prevention are, therefore, two major components that continue to underpin the pre-emptive project against identity fraud, a project that Whitson and Haggerty (2008: 574) dub as 'the *care of the virtual self'* whereby 'citizens are encouraged, enticed and occasionally compelled into bringing components of their fractured and dispersed data double into patterns of contact, scrutiny and management'.[23] And given the ubiquity of personal data and information, this project does not address only specific groups but seek to involve all citizens into the precautionary approach towards identity theft and fraud by portraying everyone as a 'potential victim' and highlighting the vulnerability stemming from exposure to and performance of routine activities. As the following statements indicate:

> Research from Experian's CreditExpert.co.uk service reveals that fraudsters across the UK are turning their attention away from wealthy individuals to people with lower incomes who rent their homes. This is probably because their personal details are easier to steal. People who rent their homes frequently share hallways or mailboxes, making it easier to intercept their post. They also tend to move home more often, which can present more opportunities to fraudsters. While some people might be statistically less likely to be targeted by fraud, it can still strike anyone at anytime.
>
> (CIFAS, 2009: 6)

[A]re you a man? Between the ages of 40 and 50? Employed? Living in the South East London area? If the answer is 'yes' to all of these questions, then you are standing in the fraudster's sights. Analysis of the frauds filed to the CIFAS database shows that you chaps will be targeted and that fraudsters will – most probably – be trying to get credit cards in your name. If you did not answer 'yes' to all of the above, however, don't get too complacent! Everybody is a potential victim [...] The identity thief may be anonymous – but they are real, and they are out there: don't make yourself a target!

(ibid.: 20)

Presenting identity fraud as a problem that can strike anyone including low-income groups and as a risk that can seep into the mundane activities of everyday life, leads to the heightening and spread of anxieties surrounding this issue. Such anxieties, according to Whitson and Haggerty (2008: 577) are 'channelled towards a series of concrete behavioural expectations that promise to reduce the prospect of victimisation' in a way that makes the precautionary discourse of identity protection both 'alarmist' and 'reassuring' (ibid.): alarmist in its anxiety-inducing tone that seeks to encourage all citizens to become more responsible for the safeguarding of information concerning their identities and the protection of their data doubles, and reassuring insofar as it represents identity fraud as something that one can protect oneself against by following specific tips and implementing various institutionally ratified mechanisms into one's daily routine. For instance, '[f]ortunately, you can take simple precautions to help keep your identity safe [...] Royal Mail and VeriSign – provider of internet infrastructure services – share some tips on what individuals can do' (CIFAS, 2009: 6, 8). These tips range from toughening the privacy settings of one's profile on social networking sites to purchasing CIFAS Protective Registration,[24] adopting vigilant bodily comportments when using a bank machine,[25] shredding one's documents before throwing them away and so on. Institutions, in this sense and as Whitson and Haggerty (2008: 577) argue, are portrayed as 'wise and benevolent, concerned to impart knowledge that will steer individuals clear of the risky shoals of an information society'. However, in the actual occurrence of fraud, as the authors assert, '[v]ictims are continually reminded to operate on the assumption that nobody will work on their behalf' (ibid.: 581) so much so that the 'material costs of the initial fraud or theft of data can pale in comparison with the frustrating and time-consuming work required to rectify the problem', prove one's victimisation and restore one's 'credit' reputation (ibid.: 580–2). Ironically, then, the availability and increasing amount of advice on how to protect one's identity only end up placing extra burden on individuals while relieving the responsibility of institutions insofar 'ignorance' or 'negligence' from the part of the individual is regarded as no excuse for losing control over personal information – but as 'a lack of self-mastery – a moral failure in the individual's duty to take care of themselves – the inevitable, if not deserved, outcome of their own lack of virtue' (Marron, 2008: 30), while the incompetence or indifference of institutions is often excused on the basis of the sheer volume of identity theft and the nature of the crime, perceived as a challenge to the realistic capacities of institutions and a strain on their resources. So,

> [w]hereas most crime victims are expected to do little more than contact the police, identity theft victims are positioned as *the* agent most responsible for rectifying their situation [...] As is the case in all highly responsibilized sectors, when individuals are victimized they are subtly encouraged to blame themselves for not having adopted any number of recommended precautionary measures.
>
> (Whitson and Haggerty, 2008: 580, 589)

Being positioned at the centre of this twin process of responsibilisation-victimisation, the individual is forced to embrace the 'responsible-victim identity' whereby she feels responsible for becoming a victim and victimised by virtue of being responsible. This involves, in addition to navigating through the bureaucratic maze of various institutions, the management of a whole host of affects that emerge as part and parcel of subjectively undergoing the experience of identity theft or fraud: 'The psychological effects too are not inconsiderable, with victims reporting a variety of reactions: from fear, anger and distress, through [the] prolonged cautiousness and suspicion described by some victims as something akin to paranoia' (CIFAS, 2009: 3). Such affects do not only stem from the loss of one's sense of privacy and control over one's credit identity, but also from the difficulties of proving and maintaining one's innocence before the police and creditors that often involve, and ironically so, the disclosure of *more* personal information. For in the absence of an identifiable perpetrator (what CIFAS coins as the 'Anonymous Attacker'), it is the victim herself that becomes the main focus of institutional distrust (in the sense that every victim is perceived as a potential fraudster given the criminal-less crime of fraud) and the 'predominant object of statistical knowledge, trend predictions, risk profiling and bureaucratic dataveillance' (Whitson and Haggerty, 2008: 585) so much so that the crime itself is conceived as merely the beginning; much of the emotional trauma is the result of the arduous process of actually attempting to restore one's identity – a tribulation which can impart such feelings as anger, impotence and frustration in the individual [...] Taken together, these are adjudged by a professional psychologist to be classic symptoms of Post-Traumatic Stress Disorder (PTSD), comparable to those of serious physical assault.

(Marron, 2008: 26, 28)

Thus, partaking of this process of responsibilisation-victimisation is also a process of 'medicalisation' whereby identity theft is articulated as a form of disruption that might threaten one's sense of identity and security, and produce an array of psychosomatic problems (including sleep disturbance, depression, panic attacks, gastrointestinal problems, etc.),[26] some of which may have a long lasting impact on the individual. Accordingly, discourses of 'healing' and 'recovery' pervade the self-help guides that are made available to victims of identity theft and fraud.[27] Like in any trauma recovery programme, victims of identity crime are 'encouraged to act upon themselves under such headings as "regaining emotional balance", "overcoming feelings of powerlessness" and "moving into activism"' (Marron, 2008: 28). Managing identity, in this sense, is not only a matter of encouraging rational capacities in citizens, but also a matter of inculcating the needed skills for managing the emotional responses and attitudes towards the risk and occurrence of fraud. What is particularly interesting and relevant to our discussion is how the array of affects and emotions relating to identity crime is itself mobilised through governmental and agential strategies and discourses as a way of making the management of identity and the care of the virtual self an even more urgent individualised task. In such strategies, victims of identity theft and fraud serve as 'cautionary examples' to others. Their dramatic testimonies are deployed as a means of giving flesh to the rationalities and motives of identity securitisation. For instance, in the CIFAS report, an entire section under the title 'Experiencing Fraud' is devoted to victims' testimonials. For instance:

Mr McKennna

I had no idea when or where or how exactly the fraudster got my card details, and because of that I am still cautions [...] I remember that I actually felt sick when I found out my account had been taken over again [...] it felt as though my privacy and security had in some way been violated. I'm not sure there is a price tag for that, is there?

(in CIFAS, 2009: 10)

Mr Ash

Mathew Ash's experience of being a victim of identity fraud left him shaken. 'I have to be honest, I do get angry' he notes. 'it's hard to quantify the effects but it has been very scary to know that someone has the majority of my personal details and will probably use them again.'

'The reason my score had gone down was due to multiple applications being made in my name, at my address, in a short space of time. So I could not obtain a loan because the fraudster had been applying on my behalf and my credit score dropped.' Mr Ash then went through the associated stress of cleaning up his credit file and removing data that had been caused by the initial fraudulent applications.

It seems strange to hate someone you don't know and will probably never meet: but then you think you may know them and you worry that it's someone close. You do get paranoid and I just hope that's the end of it, to be honest.

(ibid.: 10–11)

Such affective accounts give us a glimpse of what is at stake for the neoliberal citizen in her encounter with the experience of identity fraud. Their inclusion in the CIFAS report serves to amplify existing anxieties about this issue while at the same time promoting the proposed identity management techniques and technologies as pre-emptive mechanisms for reducing the risk of fraud and preventing its adverse emotional effects. As we mentioned before, following on from Isin's assertions, the neoliberal subject is neurotic as much as it is rational and so, governing through affects is just an important aspect of neoliberalism as is governing through freedom. Governing through affects, like governing through freedom, also raises the question of trust, and when it comes to managing and protecting identity, citizens are warned against 'taking too much on trust' (CIFAS, 2009: 22) and encouraged to change the way they think about their identity:

in order to safeguard our identities, we need to start treating our identity in the same way that we treat our property or our private lives [...] before we answer just any questions asked of us, maybe we need to remind ourselves first to pause, and think: 'Who precisely is asking me? What details are they asking me for? And why are they asking me?'

(ibid.)

In a sense, such modality of governing is underlined by, and is itself a contributor to the proliferation and normalisation of, a culture of mistrust that is increasingly infiltrating every level of neoliberal inter-actions: from the macro relationships between government and citizens and between citizens and institutions to the micro relationships between one citizen and another. This, particularly given the anonymous nature of the 'fraudster' and the criminal-less aspect of identity crime, which render every citizen as both a potential victim and a potential suspect. At the same time, governing through affects also involves 'tranquilising' (Isin, 2004: 228) the anxieties of the consumer-citizen so that the individual, 'in addition to maintaining a strict vigilance over their identity, must also know "when to relax" and "when to raise the red flag"' (Marron, 2008: 32). It is in such a context that the promotion of biometrics and identity systems as a solution to the problem of identity theft and fraud can also be seen as a 'tranquilising promise' which, while partaking of the management of unease, also attempt to appease the anxieties surrounding this issue by claiming to offer the technological fix, the panacea for identity crimes. It thereby imbues biometrics and identity systems with more value, significance and justification, investing them with emotional qualities rather than merely rational ones. Like freedom, neurosis too manages to provide a vehicle for legitimising control and a platform upon which citizens can claim their individuated rights to tranquillity and security. Biometric citizenship is a neurotic citizenship to the extent that it is embedded within and feeds upon this culture of mistrust, suspicion and neurosis, whereby various affects and emotions are continuously induced and reproduced, and where fears and anxieties are rendered as the object of management as well as a means for government.

Conclusion

In this chapter, we have addressed the issue of the securitisation of identity through biometric technology by drawing on various examples including that of border management and the phenomenon of identity theft and fraud. One major question that has been animating our discussion is: what kind of citizenship is the biometric citizenship? Through the lens of the governmentality thesis, we argued that biometric citizenship is at once a neoliberal citizenship, a biological citizenship and a neurotic citizenship. Each of these formulations provided us with distinct but interrelated visions as to what is at issue in reconfiguring citizenship through biometric technology. Importantly, what these formulations have in common is a sense of 'disjoining' and 'thinning' of the very ideal and practice of citizenship whereby the latter is becoming increasingly reduced to processes of identity management and the implementation of technical operations. Placed in context with the previous chapters, and in juxtaposition with the figure of the asylum seeker, we can see how, for the figure of the citizen, the securitisation of identity through biometric technology yields different material experiences and is imbued with a different symbolic order, revealing the polysemic, multifaceted and context-specific aspects of identity securitisation: for the citizen, identity cards, for instance, can represent a token of belonging to a national identity (being British, for instance), and biometrics can facilitate the exercise of a surplus of rights (as with the case of the IRIS system) and constitute a tranquilising mechanism for soothing anxieties relating to the risk of identity theft and fraud.

It is not, however, a matter of positing a dichotomy between the figure of the citizen and that of the asylum seeker as if they were alien to one another. For as we have argued throughout this book, these figures are co-constitutive. Nor is it a matter of simply suggesting that what is at issue is merely the problem of exclusion that can be remedied by fully including non-citizens into a so-called 'community of citizens' (Schnapper in Balibar, 2004: 69) and allowing them to equally access the advantageous sides of biometric citizenship – as this would uncritically imply that achieving the status of the neoliberal citizen is the ideal that everyone should strive for. It is true that 'securitization contributes directly to the intensification of conventional citizenship practice, as biometric technologies are employed to conceal and advance the heightened exclusionary and restrictive practices of contemporary securitized citizenship' (Muller, 2004: 279). It is also true that, as Nyers (2004: 204) argues, '[t]he ensembles of relationships that constitute state societies have gone global [...] in a highly uneven fashion: some are taken along for ride, while others are disregarded; some reap great benefits from global life, while others suffer from profound exploitations'. Yet even if those left behind get to join the ride, this will hardly solve the deeper problems lurking beneath the processes of neoliberalisation and the conception of citizenship itself. For after all, exclusion takes place also in the inside rather than merely on the outside, and contemporary enactments of citizenship ideals seem to leave no one immune to their attendant problems and challenges: while the excluded (or inclusive excluded) are made to endure the symbolic and material violence of abjection and their continuous casting as unsavoury and dangerous agents of global insecurity (Nyers, 2003: 1070), the included (neoliberal citizens), on the other hand, have to endure the neuroses, fears and anxieties that are part and parcel of living by the rules and with the ethos of neoliberalism. Inclusion is thus just as problematic as exclusion. The challenge, then, remains with the very notion of citizenship itself which, and as per Agamben's above statement, has increasingly become less about the active participation in the public sphere and more about the enactment of various technological transactions and procedures, and whereby the population is imagined less as a political community and more as a biomass entity. For such reasons, Balibar (2004: 69) is right in asserting that

> We cannot be content simply to reiterate the sort of generic discourse on inclusion and exclusion, inside and outside, belonging and nonbelonging that underlines the invocation of a 'community of citizens.' In a conjuncture marked by both the appearance of new practices of exclusion and by a vacillation of the borders of the community (or by a profound an durable uncertainty as to the type of political institution that will be able to serve as a guarantor of universal access to citizenship), we need to rethink the antinomies that are at the base of the very notion of 'community.'

In the next chapter, we shall take up this very task: rethinking the notion of community itself. Such a task requires us to regard community not simply as a 'project', an ensemble of practices and activities as is often perceived throughout the governmentality approach, but also as a broader set of 'meanings' whose underlying assumptions and beliefs continue to play a pivotal role in acts of exclusion, securitisation and so on. It requires probing into and questioning the very foundational categories (e.g., citizen, subject, other, sovereign and so on) that have been informing Western politics and its approach towards identity and belonging. To help with this task, we shall draw upon Jean-Luc Nancy's *ontological* take on the question of community and that of the political. The aim here is to emphasise the role of *relationality* and *singularity* (invoked in Chapter 3) in rethinking the notion of citizenship along more ethical, more inclusive and less technocratic ways.

5
Rethinking Community and the Political through Being-with

Putting the notion of citizenship into question has led us to raise the question of community itself, for the two remain inextricably intertwined. And like citizenship, community is also a highly complex and aporetic concept. For some, it is a source of hope and an antidote to increasing individual isolation. For others, it is a term deserving of suspicion and one that carries with it the stains of a violent past. But despite, or perhaps because of, the divergent reactions that the notion of community invokes, it continues to be an important and contentious theme in contemporary political debates and governmental strategies. As discussed in the previous chapter, the reinforcement of 'common' values and habits, the strengthening of social cohesion and the integration of migrants and newcomers into a presumed way of life are all recurrent aspirations that pervade current proposals of citizenship and immigration reform as well as some of the discourses promoting recent identity systems. The notion of community is often inserted into these debates at the same time when 'individualising' ethos and practices are being encouraged within the very same context. These articulations are expressive of a particular way of understanding community, one that is based upon the desire to construct community as unity and conformity; that is, a community founded on consensus where assimilation is often presented, implicitly if not explicitly, as the cost of membership, while, at the same time, maintaining a culture of autonomisation required for market-based purposes. For many obvious and non-obvious reasons, this vision of community is not only unrealistic but also threatening. Threatening insofar as it risks overriding difference and diversity with sameness and homogeneity, and unrealistic insofar as it cannot satisfy the demands of multiplicity and hybridity that are hallmarks of contemporary societies. This vision goes, in fact, hand in hand with the culture of fear and suspicion towards otherness and for which biometrics identity systems are mobilised as a security solution. So if, as previously argued following Agamben and Balibar, what remains at stake in the current technocratic configurations of citizenship is the notion of community itself, and if the political configurations of community that are currently on offer are equally problematic given their assimilationist aspects, how can we then reconceive of the notion of community beyond both extreme individualism and stifling assimilationism, and beyond politics of control?

Of course, much has been written on the different visions and versions of the concept of community. Without reiterating this writing here, suffice to recall Secomb's (2000: 136) inventory in which she argues that while the

> diverse views [on community] are clearly distinct, they are all committed to an ideal of community founded on unity, consensus, and commonality. For Hobbs community is unified through the rule of the sovereign, for liberalism a consensus is expressed in a mythical or hypothetical social contract, the Hegelian view suggests a unification of individual and community, and the communitarian view subordinates the individual to the common will of the community. Commonality, consensus, and harmony are the common thread running through these various perspectives, though how this concordance is to be achieved is disputed.

In this chapter, we shall address a different vision of community, one that destabilises much of the above-mentioned versions and seeks to open up a space for rethinking otherwise. This vision is based on Jean-Luc Nancy's ontological approach towards the notions of the common and the political in which he challenges the familiar accounts found in modern philosophy and political theory.

The purpose of invoking Nancy's work in the context of biometric identity systems is twofold. First, it is a way of questioning, at a deeper and broader level, some of the basic categories and self-evident assumptions that underlie Western politics with its technology-driven strategies and fearful attitudes towards otherness, articulations of which are evident in the various identity schemes and in the increasing deployment of biometrics. Second, it is also a way of reflecting back upon and challenging the epistemological foundations of the governmentality approach itself. For while governmentality studies have done well in revealing how notions such as citizenship and community have functioned as a 'project' whereby individuals and populations are made up and governed, they do not go as far as to question the wider and profound ethical and political implications of conceiving these *as* a project and of functionalising 'being itself'. And in doing so, they tend to inadvertently subscribe to the very same assumptions they seek to expose. To this end, and as we shall see, Nancy's take on the notion of community provides in some way a means of overcoming this particular limitation in the governmentality approach, by questioning the 'operative' aspects of current forms of community and by bridging between the ontological and the normative, between the relational and the political.

In what follows, I begin by highlighting the ways in which Nancy's critique of available conceptions of community relates to current policies of immigration and border control. I then go on to examine Nancy's reformulations, specifically in terms of the notions of 'being-with' and 'being-in-common'. Finally, I reflect on how Nancy's abstract ideas can provide helpful directions for more concrete conceptualisations of community and politics.

Immanentism

> The gravest and most painful testimony of the modern world, the one that possibly involves all other testimonies to which this epoch must answer (by virtue of some unknown decree or necessity, for we bear witness also to the exhaustion of thinking through History), is the testimony of the dissolution, the dislocation, or the conflagration of community.
>
> (Nancy, 1991: 1)

These trenchant opening lines of Nancy's (1991) *The Inoperative Community* carry the echo of the burning concerns revolving around the notion of community that have preoccupied much of contemporary political thinking throughout recent decades. They also echo the *limit* of such thinking, a limit that has historically manifested itself, in different and at times dangerous ways, in both communitarian and liberal approaches to the question of community. Nancy's commitment to rethink this insufficiently thought out question, as Devisch (2000: 240) puts it, is not merely an attempt to labour at that limit so as to *stretch* its contour and reach out beyond such thinking, but it is more so a commitment to develop a radical *alternative* view altogether of what community *is* and what it *ought* to be. What motivates Nancy's enquiry is primarily his trepidation over contemporary views and articulations of community and politics, which he sees as often being driven by nostalgic yearnings for a lost community[1] (notably the case with the communitarians of the Right) and partaking of 'substantialist' and 'immanentist' metaphysics. By substantialism, Nancy refers to the essentialist assumption that individuals are *pre-constituted* prior to entering into a community and relating to one another, an assumption that is predominant in traditional liberal views of community and political conceptions of subjectivity.[2] And immanentism is Nancy's term for the conception of a community that is based on *self-enclosure, self-identification* and the gathering around a *common* substance and identity. It functions, in Nancy's writing, as a surrogate word for totalitarianism, designating a form of relating and governing that is not only characteristic of totalitarian states or regimes but one that represents 'the general horizon of our time, encompassing both democracies and their fragile juridical parapets' (1991: 3).[3] Thus for Nancy, the initial task at hand is very much a matter of challenging the *horizon* itself (ibid.: 8) in order to imbue our thinking of community with new and different conceptions and formulations that are not informed by the 'mirages of an origin' (ibid.: 11) and the nostalgic longings for a lost communal experience whose immanence can be recovered, nor by substantialist understandings of individuals and communities. It is, in short, a matter of rescuing the notion of community from the twin dangers of fusion and extreme individualism.

While Nancy does not directly speak of the question of immigration or citizenship when addressing the concept of immanentism,[4] there is a strong sense in which the logic of the latter pervades contemporary immigration and citizenship policies, unfolding at the level of various *figurations*. So before moving to discuss Nancy's alternative view of community, it is worth elucidating this unfolding which, as I see it, is manifested along three interrelated axes, namely *bordering*, *technicism* and the notion of *mythical collective identity*.

As we have seen throughout the discussions in the previous chapter, there has been a myriad of mutations in the way citizenship is conceptualised and performed. These mutations are increasingly challenging the idea of a bounded nation-state citizenship in which rights, entitlements and responsibilities are territorially framed. Yet, and as we have shown through the example of *The Path to Citizenship* and in previous instances, traditional notions that are based on nationality and a common set of cultural values are still upheld within the political rationalities and imaginaries. So denationalisation and renationalisation work in tandem within contemporary forms of citizenship governance. Referring to the work of Saskia Sassen, Stasiulis (2008: 136–7) points out a similar argument in the following summary:

> [T]ransformations in citizenship resulting from globalization combine partial *denationalizing* moves, whereby the growing articulation of globalization with national economies hastens the withdrawal of the state from various spheres of citizenship entitlement, with moments of *renationalizing*, as states securitize and harden their borders to the entry of illicit and poor migrants.

The amalgamation of these denationalising and renationalising moves has resulted in the subsistence of older forms of exclusion and the creation of new ones. As Balibar (2004: 76, 68) rightly argues, 'every institution of citizenship involves the institutionalization of exclusions, following different modalities [...] the logic of exclusion has changed in method as often as it has changed in historical space'. In the contemporary era, policies of citizenship and immigration have led to the emergence of forms of 'inner exclusion' (or what we previously referred to as 'inclusive exclusion'⁵ – the inclusive exclusion of asylum seekers and the waiting-to-become citizens, for instance) as well as 'outer inclusion' (engendered through flexible citizenship rights for the kinetic elite). With this, and as we argued before, liberal and illiberal practices, democratic and totalitarian forms of governing become inextricably mixed. In fact,

> the constitutive movement that gives it [the institution of citizenship] its democratic power is the same movement that carries an institutional schema of inclusion and exclusion (the institution of a 'border' of citizenship) *beyond itself* once the status quo turns out to be untenable, except at the cost of a reinforcement of police practices, and thus of violence, or cycles of violence and counterviolence, at first on the 'margins' of public space, and finally in its center.
>
> (ibid.: 76)

The *bordering* of citizenship is one of the quintessential features of immanentism insofar as it serves the function of demarcating the *vacillating* lines between the included and the excluded (at times with violent consequences) and *enclosing* the 'community of citizens', as it were. It is a process that is fed by the substantialist fantasy of autonomy and self-sufficiency, and a rejection of the porousness of the world: the world as a place of 'exposure' where 'there has to be a *clinamen*[:] an inclination or an inclining from one toward the other, of one by the other, or from one to the other' (Nancy, 1991: 3). The rationality of bordering ignores this logic of *clinamen*. It ignores the logic of relatedness. Instead, it lends itself to the logic of immanentism whereby *being-with*, *being-in-common*,[6] or in fact, *being* at all, are reduced to the *organisation* of sameness (*immanentist politics*) or the sharing of *common* substance (*immanentist community*). Immanentism, in this sense, functions at the level of self (state/demos/individual)-enclosure, that is, the sealing of the inside from the outside in order to exclude any 'unwanted' element that might permeate it. Border control and immigration policies provide the concrete framework for immanentism. That is not to say, however, that the possibility of exposure is entirely eliminated. Instead, exposure, in immanentism, becomes that which relates to exteriority only in terms of exchange value and flow of capital (in fact, this kind of exposure, as discussed before, is encouraged as it sustains the doctrine of free market and perpetuates advanced capitalism) as well as through emerging modes of measurement such as quotas for asylum seekers and the points system for work permits and residence. And here, measure is not only the quantifying of dimensionality (how many asylum seekers and immigrants should be let in) – although this is often presented in some political discourses as the salient point, but also the quantifying of 'responsibility' (Nancy, 2000: 180) so much so that the question becomes not only 'how many?' but 'which?' (Which (skilled/needed) immigrants should be given the right to enter and reside? Which asylum seekers are 'genuine'? Which asylum seekers should one be responsible to? And so on). In such a context, numbers become imbued with a 'moral magnitude' (ibid.), they become metaphors for dignity, and measure becomes concurrently the embodiment of exposure as well as enclosure, both of which are, nonetheless, operated within the intentionality of bordering and division.

This process of bordering rests upon the investment in technology, the second figure of immanentism. As seen so far, biometrics is becoming a prominent means by which borders are controlled and bodies are scanned in order to establish their (il)legitimacy and prevent the intrusion of the 'unwanted'. As with the case of identity systems, biometric mechanisms of control are becoming too pervasive, too *immanent* that borders are no longer constituted around the 'physical' alone but actualised in the everydayness of life activities and through the ubiquity of information networks. The fact that technology is an aspect of immanentist politics, is an attestation to the way in which the political itself is increasingly fading into a state of technicism (Coward, 1999: 18) – a depoliticisation of society whereby government policies and debates are merely technical discussions on the type of techniques to be deployed in order to protect borders, filter movements, eliminate infiltrations and sustain control by means of measurement and exclusion. Everywhere, according to Lacoue-Labarthe and Nancy (1997: 126–7), is 'converting itself into a form of banal management or organisation'. And it is in this organisational banality that Nancy sees the threat of immanentism. For wherever there is routinisation, technologisation and excessive management, there is also a risk of 'a more insidious and (as one says of some technologies) "softer" form of totalitarianism' (Lacoue-Labarthe and Nancy, 1997: 128): a totalitarianism that might not necessarily be experienced as 'a renaissance in its old forms, but could assume new figures latent in market democracy itself' (Hutchens, 2005: 129). As discussed in detail in Chapters 2 and 4, technologies of border and immigration control are clear examples of the paradoxical aspects of contemporary forms of biopolitics whereby processes of totalitarianism and democracy are concurrently and continuously brought into play. The function of such technologies is thus a strong epitome of immanentism.

In addition to bordering and technicism, the notion of mythical collectivised identity is also a crucial aspect of immanentism and one that is often mobilised in order to justify, articulate and sustain the function and objectives of immigration and citizenship policies. In immanentism, the idea of collective identity is bound to the idea of 'common substance' and is always represented as the essential bond between people and the foundational character of communal identification. Common substance, as such, becomes the logic of institutionalisation in immanentist politics which sees itself as the organiser and guarantor of common identity. Yet the realisation of this communal identity takes place only at the level of 'articulation' (Nancy, 1991) where the notion of 'common substance' is made 'immanent' to the idea of communality so much so that it is never questioned but always taken-for-granted and perceived as 'common sense'. And to question common sense/common substance is to put at stake the very project of immanentism and expose the inside to the irreducible outside. Immanentist politics, as such, performs its 'enclosing' task by means of suppressing/reducing difference, regulating alterity and securitising its immanentist community (all being manifested in immigration control). And when the other is 'needed' (skilled migrants/'Sector Based Scheme' migrants) or 'imposed' (having to grant access to asylum seekers because of the signed international conventions), there is a tendency to enforce modes of assimilation – what is also euphemistically called 'integration' – so that the otherness of the other is absorbed into a homogenous totality in which its 'imagined' disturbance/threat is reduced if not eliminated. For instance, 'hard work, determination and a willingness to integrate propelled them [immigrants] forward [...] Britain has an enviable record of racial integration' (Howard, 2005).[7] Or again:

> [o]ur reform of the path to citizenship is an important part of this work [integration and cohesion]. The key feature of the proposed system is that it aims to increase community cohesion by ensuring all migrants 'earn' the right to citizenship and asks migrants to demonstrate their commitment to the UK by playing an active part in the community.
>
> (Home Office, 2008e: 12)

Integration, in this sense, becomes a *work*, an achievement to be extolled as the virtue of 'good citizens' and 'good governments', all, while invoking principles of common substance and essential unity: 'That's what makes us so proud to be British' (Howard, 2005). 'To be British' is, in fact, a testimony of how the myth of communal essence speaks through the political enterprise of immanentism and renders identity as a *project*, as the gathering together of absolute figures (citizens, state, institutions, communities, etc.) in order to *naturalise* the mythical character of collective identity and sustain its myth of *absolute particularity*.

> In myth, [...] existences are not offered in their singularity[8]: but the characteristics of particularity contribute to the system of the 'exemplary life'.
>
> (Nancy, 1991: 78)

The articulation of absolute particularity and collective identity is therefore dependent on mythic, inaugural figurations that circumscribe commonality and enable the discursive process of separation and enclosure. For in immanentism, it is myth that constitutes the 'common' through naturalisation. It is myth that infuses the utopia of fusion and assumes the role of 'origin' and the founder of pre-communal time (ibid.: 50). And it is through and around myth that the 'gathering' takes place, and that the genealogy of community is established. As such, one might compare political speeches, assemblies, campaigns and so on to a scene of gathering in which the myth is being recycled by invoking the supposed genesis of absolute figures: how they came to be together and how they must protect their 'origins' and 'communal essence' from the intrusion of the outsider. In (political) speech, the articulation of myth takes place when a series of shared values and beliefs are amplified in an attempt to ennoble that speech (ibid.: 48), substantiate immanentism and present collective identity as an absolute figure whereby citizens and state are situated within an enclosure. In such a process, myth transforms its mythic status into a natural one to the extent that it is no longer perceived as a myth but becomes the condition *par excellence* for belonging, politics or any other form of 'communitarian fulfilment' (ibid.: 69).

From the above, then, we can see how immanentism figures in immigration and citizenship policies and discourses, underlying modes of inclusion and exclusion. Importantly, immanentist figurations are always presented as a 'project'; in the guise of a *work* to be *accomplished* (Nancy, 1991), be it in terms of preserving an absolute separation through bordering, the mobilisation of technology to do so, or simply the perpetuation of the mythical collectivised identity. Such figurations are problematic not only given their exclusionist aspect but also in the way they define the parameters of inclusion in terms of substantialist understandings and nostalgic longings whereby identity and belonging are reduced to and burdened by the illusive belief in a fixed common substance, a need to sustain a state of self-enclosure and an imperative to adhere to a certain pre-established set of norms and communal values. To this end, Nancy's deconstructive approach to the question of community is, above all, an attempt to *unwork* the workings of immanentist figurations in order to *open* up a different horizon for rethinking the meaning and *raison d'être* of community beyond the fusional and totalistic formulations of communitarianism and the loose and ephemeral associations of liberalism.

Being-with and being-in-common

Community is what takes place always through others and for others. It is not the space of the *egos* – subjects and substances that are at bottom immortal – but of the *I*'s, who are always *others* [...] It is not a communion that fuses the *egos* into an *Ego* or a higher *We*. It is the community of *others*.

(Nancy, 1991: 15)

To want to say 'we' is not at all sentimental, not at all familial or 'communitarian.' It is existence reclaiming its due or its condition: coexistence.

(Nancy, 2000: 42)

Nancy's entire vision regarding the notion of community is at once notoriously abstruse and strikingly simple,[9] highly philosophical and humbly empiricist.[10] While not being necessarily systematic, it does fellow a series of threads whose origins reside with a variety of thinkers, ranging from Hegel, Nietzsche and Rousseau to Derrida, Blanchot and Bataille, assuming a rather hybrid and multidimensional aspect. One of the key tenets of Nancy's approach towards the notion of community is the Heideggarian question of being-with (*Misteinsfrage*). This question, as its composition suggests, is about sociality, about the role and the status of the individual being with regard to social relations. It follows from Heidegger's (1962) assertions that the self is formed in relation to others and to its ontological position of being-there (*Dasein*). As per Devisch's (2000: 241–2) cogent elaboration:

Since I have been thrown into the world, my entrance into sociality is not an independent and solipsistic decision. I am already inscribed in the world even before my self-sufficient will was able to decide to do that. The ontological fact that I am already a social being prevents me from dreaming of myself as an independent being who is free to enter into sociality and who gives sense to his own life. To be thrown into the world implies that I am, as a *Dasein*, co-original with a *Mitsein* [...] There is no isolated I without others. The comprehension of others is always implicated in the comprehension of *Dasein* and its being-in-the-world.

This stance, which can be summarised as 'I am with therefore I am', indicates the importance of the 'with' and its co-originality with *Dasein* in Heidegger's thesis. Yet, for Nancy, this question of being-with remains rather underdeveloped in comparison to Heidegger's elaborations of the question of Being (*Seinsfrage*):

In twentieth-century philosophy, the Heideggerian ontology of *Mitsein* is still no more than a sketch [...] Even Heidegger preserves

this order of succession in a remarkable way, in that he does not introduce the co-originarity of *Mitsein* until after having established the
originary character of *Dasein* [...] It is just as much a question of
doing justice to the essential reasons for why, across the whole history of philosophy, being-with is subordinated to Being *and*, at the
same time and according to this very subordination, is always asserting [*de faire valoir*] its problem as the very problem of Being. In sum,
being-with is Being's own most problem

(Nancy, 2000: 31–2, 44)

As such, Nancy makes it his task to develop a 'co-existential analytic'
(ibid.: 93–9) where *Mitseinsfrage* would be equi-primordial with (if not
even more primordial than) *Seinsfrage*, that is, an analytic where the
concept of Being is always *already* a being-with, understood in terms of
finitude, singularity, relationality and sharing. This emphasis on *Mitsein*,
as Caygill (1997: 22) puts it, 'signals a move from a thinking of being
as substance'. It constitutes a challenge to past metaphysical proposals
and philosophical speculations in which Being is seen as pre-existing
and preceding the possibility of being-with others. In this way, Nancy
is performing an *inversion* of 'the order of ontological exposition' as a
means of rethinking how we understand our being-in-the-world and
stressing that the 'with' is not simply an adjustment of *Dasein* or an
addition to being, but is at the heart of Being itself (ibid.: 30–1). Ultimately, the question, for him, becomes not so much 'how we might
establish a bond between us, but rather [how] it is that we have come
to consider ourselves separate in the first place' (Edkins, 2005: 383). So,
'at the point where Heidegger lost the thread, Nancy picks it up again
to knit the "question of being" together with the "question of community"' (Devisch, 2000: 242). And what drives this task is not merely the
desire to expose the philosophical shortcomings vis-à-vis the question
of being-with and community but also to turn away from Heidegger's
defective nationalistic pathos in which being-with is grafted onto a
'destiny' and a 'people' (ibid.):

All of Heidegger's research into 'being-for (or toward)-death' was
nothing other than an attempt to state this: *I* is not – *am* not – a subject. (Although, when it came to the question of community as such,
the same Heidegger also went astray with his vision of people and a
destiny conceived at least in part as a subject, which proves no doubt
that Dasein's 'being-toward-death' was never radically implicated in

its being-with – in *Mitsein* – and that it is this implication that remains to be thought.)

(Nancy, 1991: 14)

Saving the 'with' from the oppressive and devouring power of a 'We' and a 'Subject' is thus a prerequisite task if the question of community is to be rethought on the basis of being-with and in a non-immanentist way. So what does the 'with' stand for anyway? Certainly, for Nancy, the with has nothing to do with substance, gathering, communion, fusion, or the aggregation of egos. It is neither *presentable* nor *unpresentable*. Rather, the with is

> the mark of unity/disunity, which in itself does not designate unity or disunity as that fixed substance which would undergird it; the 'with' is not the sign of a reality, or even of an 'intersubjective dimen-sion.' It really is, 'in truth,' a mark drawn out over the void, which crosses over it and underlines it at the same time, thereby consti-tuting the drawing apart [*traction*] and drawing together [*tension*] of the void. As such, it also constitutes the traction and tension, repul-sion/attraction, of the 'between'-us. The 'with' stays between us, and we stay between us, but only [as] the interval between us.
>
> (Nancy, 2000: 62)

Simply put, the with is a dynamic movement that brings us together without gluing us with each other, that draws us apart without sepa-rating us forever. It implies *connection* as opposed to fusion, *exposure* as opposed to communion, a togetherness that is 'neither the sum, nor the incorporation' (ibid.: 33) and a separateness that is neither atomistic nor individualistic. It *is* the condition of every singular being, the condition of its co-existence, its *co-ipseity*, its co-appearance[11]: its being *singular plural*. *Ego sum expositus*. Thus is the ontological structure of Being, according to Nancy. So, before assuming any identity, beings are already *exposed* to a space in-common and *open* to an act of sharing where what is being shared is not a substance or a pre-given identity (be it individ-ual or communal), but the *nakedness* of being-with-one-another and the *experience* of sharing itself. This is

> the plural singularity of the Being of being. We reach it to the extent that we are in touch with *ourselves* and in touch with the rest of beings. We are in touch with ourselves insofar as we exist. Being in

touch with ourselves is what make us 'us,' and there is no other secret to discover buried behind this very touching, behind the 'with' of coexistence.

(ibid.: 13)

And this is precisely the horizon out of which the question of community needs to be rethought, according to Nancy.

With these ontological reconfigurations, Nancy is challenging both traditional conceptions of community and liberal notions of individuality by staging a confrontation between 'common-being' and 'being-in-common', between individuality and singularity. At one register, he argues that community is not

a project of fusion, or in some general way a productive or operative project – nor is it *project* at all (once again, this is its radical difference from 'the spirit of a people,' which from Hegel to Heidegger has figured the collectivity as project, and figured the project, reciprocally, as collective).

(Nancy, 1991: 15)

So far from being a work of identitarian appropriation, a gathering under a common-being in which 'a pre-given or pre-supposed identity or substance (in the form of a people, a nation, a class, etc.) crystallizes itself into a figure, name or myth' (Devisch, 2000: 246), community, for Nancy, is about *being-in-common* where the 'in' does not stand for a social or an economic bond that ties one already-given subject to another, but a differential spacing between singular beings that 'consists in the appearance of the *between* as such: you *and* I (between us) – a formula in which the *and* does not imply juxtaposition, but exposition' (Nancy, 1991: 29). As such, '[t]he "in" makes of community a verb, not a substance as the work of a pre-existing communal essence' (Devisch, 2000: 252): a verb that primarily denotes an act of sharing and a process of *compearing* (or co-appearing).

The being of community, in this sense, is the exposure of singularities in/to their being-in-common, and community is the name of this *dynamic* and *ongoing* relational experience of exposure rather than a *fixed* project, an operation, a structure or an organisation: one does not produce a community, one experiences it and is constituted by it. Community is always a *non-work in progress*. 'We cannot appropriate the *in*. It is sociation, the spatiality of our Being-in-the-world, of our Being-exposed-to. The in-common is not a mere modification of our being.

It means that we cannot exist without being exposed to others, without coexisting. We are as Being-in-common' (Devisch, 2000). Again, being is being-with. This is why Nancy vehemently objects to the discourses of loss and recovery that permeate philosophical and political constructions of community. He insists that nothing has been lost and hence nothing needs to be recovered insofar as community, 'far from being what society has crushed or lost, is *what happens to us* – question, waiting, event, imperative – *in the wake* of society' (Nancy, 1991: 11). It is therefore a grave mistake, if not even a dangerous move, according to Nancy, to strive for recuperating the immanence of some ideal model of community that purportedly existed at some point in history – as this would lead to immanentist, parochial and stifling understandings of the idea of community that are antithetical to Nancy's consideration of community as a mutual and spontaneous compearing within an open and singular space of sharing which is neither reducible to a common identity nor to a specific goal – a *telos*. Community takes place singularly in its own taking place, and this taking place is unrepeatable and unworkable; it cannot be objectifiable or reproducible '(in sites, persons, buildings, discourses, institutions, symbols: in short, in subjects)' (ibid.: 31). Coexistent with this taking place is, indeed, a potential encounter that has no structured individualising or universalising principles, that knows no border, no substance, no limit (except that of birth, death and alterity), no point of departure, no point of arrival. It imbues itself with life – as well as death – at the dynamic moment of crossing of ways, at the time of the *interim* and within the space of the *in-between*.

It is with equal vehemence that Nancy also attacks the idea of neoliberal individualism. To him,

> the individual is merely the residue of the experience of the dissolution of community. By its nature – as its name indicates, it is the atom, the indivisible – the individual reveals that it is the abstract result of a decomposition. It is another, and symmetrical, figure of immanence: the absolutely detached for-itself, taken as origin and as certainty.
>
> (Nancy, 1991: 3)

This is why it is crucial not to conflate singularity with individuality. For if one overlooks the distinction between the two, one will have overlooked the entire edifice of Nancy's critique. For Nancy, singularity is not bound with the atomistic character of individuality. 'Singularity never has the nature of the structure of individuality. Singularity never takes place at the level of atoms, those identifiable if not identical identities; rather it takes place at the level of the *clinamen* [inclination], which is unidentifiable' (ibid.: 6–7). It is therefore always implicated in a relationship of sharing and exposure. The singular is plural:

> The concept of the singular implies its singularization and, therefore, its distinction from other singularities (which is different from any concept of the individual, since an immanent totality, without an other, would be a perfect individual, and is also different from any concept of the particular, since this assumes the togetherness of which the particular is a part, so that such a particular can only present its difference from other particulars as numerical difference). In Latin, the term *singuli* already says the plural, because it designates the 'one' as belonging to 'one by one.' The singular is primarily *each* one and, therefore, also *with* and *among* all the others [...] The togetherness of singulars is singularity 'itself.' It 'assembles' insofar as it spaces them; they are 'linked' insofar as they are not unified.
>
> (Nancy, 2000: 32)

So, if community is the experience of being-in-common where the common is neither a substance nor an identity but the space of sharing between singularities, what does become of the status of the political?

> The political, if this word may serve to designate not the organization of society but the disposition of community as such, the destination of its sharing, must not be the assumption or the work of love or of death. It need neither find, nor regain, nor effect a communion taken to be lost or still to come. If the political is not dissolved in the sociotechnical element of forces and needs (in which, in effect, it seems to be dissolving under our eyes), it must inscribe the sharing of community [...] To attain such a signification of the 'political' does not depend, or in any case not simply, on what is called a 'political will.' It implies being already engaged in the community, that is to say, undergoing, in whatever manner, the experience of community as communication.
>
> (Nancy, 1991: 40–1)

The political, in this sense, is not so much a matter of producing and maintaining structures around which individual and communal identities could be organised. It is not that which can be distilled from a program of 'politics' per se. For Nancy, the political is more of a *question* of relationality, communication, engagement and the spacing between singular beings, a question that cannot be reducible to traditional modes of linking between citizen-subjects nor be subsumed under a specific political project or practice. In fact, Nancy, in collaboration with Lacoue-Labarthe (1997), makes a clear distinction between politics as an *empirical* practice and the political as the *essential nature* of politics – a distinction that turned out to be one of the key points of departure for Nancy's body of work on the question and the space of the political and also a source of contention for reasons we shall mention later on. Crucial to Nancy's and Lacoue-Labarthe's endeavour is a need to save the political from being obscured by and dissolved into a totalising and pre-determined political paradigm – or rather, into a political paradigm *tout court*. They identify a sense of *withdrawal* or *retreat* of the political within the techno-social order whereby the political has become commensurable with empirical practices of politics, reaching a state of closure and 'the obviousness of an "it goes without saying"' (Lacoue-Labarthe and Nancy, 1997: 126). But this closure, according to the authors, 'opens onto "something"' (ibid.: 132). It 'makes something appear or sets something free [;] something which would be "the political" – or the essence of the political – drawn back from the total completion of the political in the techno-social' (ibid.: 131–2).

To put it simply, it is because (rather than despite of the fact that) the political has withdrawn, as a result of the dominion of the techno-social field and its political practices, that the question of the political can then be retraced, raised and thought afresh.[12] For in its retreat, the political has left a lacuna whereby the all-too-familiar *figures* of state, nation, people and the like no longer suffice for the definition of the space of the political and that of community (see also Critchley, 1999; Hutchens, 2005; James, 2006). This, however, does not mean that the lacuna has been left empty. Rather, it has been refilled, to the point of saturation, with the immanence of politics, of which 'everything is political' has become a stamp expression: the political has withdrawn; its space has been occupied by political practices which, by virtue of their immanence, has rendered everything political so much so that the political has come to mean *everything* and therefore *nothing*. The political is no longer in question.

So together, in what would seem a rather peculiar but somewhat justified move, Nancy and Lacoue-Labarthe call for a second retreat of the political, this time, a 're-treat' from the sphere of politics itself so that the political can be revitalised as a question for *philosophical* scrutiny instead of remaining as self-evident and taken-for-granted. In other words, the authors incite philosophy to take a step back from politics in order to reflect, at a distance, on what has become of the nature of the political.

This distance, nevertheless, is not meant to secure a safe haven or a comfortable zone wherein philosophers could remove themselves form the messy reality of politics (although it may seem so, at first glance and without a deep engagement with the vicissitudes of this proposal), but to create a space for the labour of *thinking* which, for Nancy, is by far the most pressing and ethical task of our time. As James (2006: 165) puts it, 'the withdrawal from politics and the concomitant "retreating of the political" is a deeply engaged gesture that does not intervene or make prescriptive/normative judgments about the present, but that demands that the present be *thought*'. To think the present is to reclaim it, to save it from the jaws of nostalgic lamentations over a conjured up past and from the fear-ridden projections towards an unknown (or rather, desired-to-be-known-in-advance-at-any-cost) future. It is the demand to be mindfully *present* in the present so as de-clutter the space of the political from calculations, programmes, projects, fears, promises, in short, from the immanence of politics, all with the aim to open up a different horizon for thinking the political. The retreat is indeed an *active* retreat:

> the gesture of the 'retreat' cannot be a simple one. It is not a matter of turning away from the political [...] No retreat, no *safe haven*, if you like, could accommodate and protect the one who 'retreats'. Which amounts to saying that the 'retreat' has to be *active*, offensive, even.
>
> (Lacoue-Labarthe and Nancy, 1997: 96–7)

Yet Lacoue-Labarthe's and Nancy's withdrawing of the political has, unsurprisingly, produced its own 'withdrawal symptoms', manifested in the criticisms directed at their attempt to formalise the distinction between the political and politics, and to place the question of the political exclusively within the ambit of philosophy and away from the empirical field. More specifically, and as James (2006: 159–60) explains, by taking this stance, the authors expose themselves to criticism on two sides.

> On the one hand there will be those involved in philosophical or theoretical reflection who will demand that they make their thinking more directly engaged with politics and the immediacy of political struggle. On the other hand there will be those who, engaging in empirically based political science, would resent or refuse the attempt by philosophy to rework or dismantle some of its grounding assumptions.

Examples of these criticisms include the responses from Fraser, Critchley and Norris whom, despite their differing perspectives, share a reservation towards what they view as an overemphasis on the ontological framing of the political at the expense of engaging with the pragmatic field of politics (see James, 2005, 2006). Likewise, Caygill (1997) also criticises Nancy's and Lacoue-Labarthe's transcendentalising of the political and their lack of reflection on the *violence* that he sees, following Hannah Arendt's analysis of *Mitsein*, as inherent to philosophical considerations of the political.

Such contentions have doubtless their own merit in pointing out some of the limitations and concerns that might transpire out of the thinking of the political in a (quasi-)transcendental way. But in doing so, they also do not allow for a total openness to the possibility of *thinking-with* Lacoue-Labarthe and Nancy, and appreciating what their alternative view has to offer. It is true that political thinking cannot do without thinking political struggle. It is true that violence is a danger that is always looming over the act of sharing, even when what is being shared is not tangible wealth or resources but the experience of sharing itself (an experience involving the expenditure of some kind of energy, after all). Yet retreating the political from politics does not necessarily imply a disregard towards political struggle or violence, nor towards attempts to resist them. It only implies a refusal of the notion that philosophy should provide *working* theoretical frameworks for political projects. As James (2006: 153) argues with regard to Nancy's work in general,

> Nancy's thought appears to be responding to two contradictory impulses. On the one hand it resonates consistently with profoundly political implications and often quite directly, and specifically, address the political in both its historical and contemporary dimensions. On the other hand, *as* thought it makes no attempt to found or endorse a specific politics or political program.

The latter point does not amount to abrogating philosophy's responsibility towards political engagement, but a refusal to operationalise and thereby immanentise the task and the movement of thinking itself. Dallmayr (1997: 191) presents a similar defence in suggesting that Nancy's work carries

> a profound normative significance: a significance resulting chiefly from his pronounced 'anti-totalizing' or anti-systemic stance. Politics

and 'the political' cannot neatly be divorced. The absence of communitarian substance does not mean a lack of bonding, just as the
accent on 'in-operation' does not entail a lapse into indifference of
apathetic inaction. Precisely the disruption or 'interruption' of total
structures carries with it a political and moral momentum.

So, instead of dwelling on the negative responses to Nancy's work, suffice, at this occasion, to make a note of them and move on to see what
constructive and helpful elements can be drawn from Nancy's reflections, particularly in relation to the overall context of this book. Before
we do that though, we need to attend to the question that is still out-
standing: now that the political has been doubly retreated (in deed and
in thought; first *into* politics through the sheer dominion of its practices and second *from* politics through the philosophical manoeuvre
of rethinking), what is, then, the essence of the political, according to
Nancy?

As mentioned earlier, for Nancy, the essence of the political is a
question of *relation*. It *is* relation rather than organisation. It is that
which takes place between singular beings in the space of their being-
in-common, unhindered by institutionalisation and unburdened by
operations, were it not that these relations are constantly under threat of
being hijacked by political practices, structures and rationalities *time and
again*. At first sight, Nancy's thought on the political seems to diverge
significantly from the Foucauldian approach of governmentality, a point
that is made explicit in this passage from *The Experience of Freedom*:

> the political does not primarily consist in the composition and
> dynamic of powers (with which it has been identified in the mod-
> ern age to the point of slipping to a pure mechanics of forces that
> would be alien even to power as such or to a point of a 'political
> technology,' according to Foucault's expression), but in the opening
> of a space.
>
> (1993: 78)

'Every relation is a power relation'. Well, it does not have to be that
way, according to Nancy's thinking. Rather, every relation is a relation
of exposure, the exposure to one another's vulnerability and finitude by
virtue of one's being-toward-death which exceeds the boundaries and
the resources of the subject and that of subjectivity[13]:

> Sharing comes down to this: what community reveals to me, in pre-
> senting to me my birth and my death, is my existence outside myself.

Which does not mean my existence reinvested in or by community, as if community were another subject that would sublate me, in a dialectical or communal mode. *Community does not sublate the finitude it exposes. Community itself, is nothing but this exposition.* It is the community of finite beings, and as such it is itself a *finite* community. In other words, not a limited community as opposed to an infinite or absolute community, but a community *of* finitude.

(Nancy, 1991: 26–7)

The political, in this sense, is nothing other than the opening of/to this community of finitude, the community of singular beings *as* finite and mortal beings. It is outside of and beyond all projects of subjectivity, (in)dividuation and identity. It stands in opposition to the dominant articulations of the political as 'the techno-economical organization or "making operational" of our world' (ibid.: 23), articulations that are based on 'an essentialized understanding of the human as *homo economicus*' (James, 2006: 186). The inoperative community, as such, 'recasts the political outside any possibility of grounding or any assumption of collective identity, and outside any possibility of project or historical process' (ibid.). It thereby offers an alternative view of what a 'political community' is, a view that does not reduce the latter into an exclusive club whose cost of entry is assimilation and subsumption under a common being and a common set of values (Schwarzmantel, 2007: 469), but opens it up to the possibility of being rethought as a space of mutual compearance, engagement, communication and the sharing of singularity and difference. A political community, in this sense, is

neither a pre-existing substantial entity [...] nor an entity that gains a substantial identity through mimetic reflection and playing back (ideology laden with platitudinal wisdoms and traditional symbols) or through reference to transcendental values ('the Chosen People', 'His Most Christian Majesty', 'God Bless America and to Hell with All the Rest').

(Hutchens, 2005: 158)

Parenthetically, there is, to be sure, a kind of affinity between Nancy and Foucault in the way they both reject the foundational figures of the state, the people, the subject, the sovereign and so on as already given templates. But how each proceeds after this rejection is rather different. Foucault (2008 [1979]) turns to concrete practices as a starting point for his analysis of governmentality. Nancy, on the other hand and as discussed above, sees in dwelling on these practices a hindrance to (re)thinking the political outside of the frame of politics. He turns instead to ontology and more specifically to the question of being-with which, according to him, is anterior to and beyond practices, and thus it cannot be appropriated by the working of politics and its technologies. As Levett (2005: 427) puts it,

> nothing precedes the opening of [the space of the political], just as nothing gathers up the dispersed beings whose exceptional openings allow its purely differential constitution [...]; each singularity becomes the place-holder of the whole 'movement' of being-in-common, just as each singularity remains unthinkable without its exposure to *other*, equally different, or even incommensurable, singularities.

In reconceiving the political in this way, Nancy is also opposing the Schmittian substantialist approach whereby '[t]he people or the nation remains the origin of any political events' (Schmitt in Levett, 2005: 424) and where the division between friend and enemy is what constitutes the essence of the political.[14]

Nancy's demand is, therefore, a demand to rethink the political beyond both foundational figures and political practices. It thereby places an exigency on every perspective, engaging with the sphere of politics, to be mindful of and reflective upon its own categories, assumptions and forms of understanding and pre-understanding, philosophically before even considering its empirical dimension. The demand is in a way a methodological and epistemological demand. Failure to do so, according to Nancy, would only lead to limiting the critical force of these perspectives, if not even unwittingly contributing to the very immanentist thinking that drives much of contemporary political practices.

The question remains as to how Nancy's ontological reconfiguration of the political can have an impact on politics and inform its concrete practices. Obviously, Nancy's approach provides no easy answer to such a question or a means of simple and practical application. As James (2006: 339–40) argues:

> Nancy's response [...] will necessarily remain disappointing to those who look to philosophy to lay a ground or foundation for political projects or decisions. This is because he resolutely refuses any straightforward movement between the order of philosophical reflection on the one hand and that of politics on the other. He

refuses the notion that philosophy should lay theoretical grounds for a project or program which would then be conceived as the concrete effectuation, or completion, of the philosophical within the realm of the political. The expectation of such a movement from philosophical reflection to political project articulates, according to Nancy, the very essence of the metaphysical attitude within philosophy, and is deeply implicated in the recent history of European totalitarianism and the destructive or genocidal energies which that history unleashed.

So, 'what is to be done?', this is in fact a question that Nancy himself asks at the end of *Retreating the Political*. But he asks it not by way of providing a ready-to-be-implemented answer, but to reveal once more the assumptions and dichotomies underlying common attitudes vis-à-vis the so-called passage from thought to action. 'What is to be done?' is indeed a problematic question, for Nancy, precisely because it presumes that we have exhausted and completed the task of thinking and we are now ready for action. This is exactly the teleological attitude that Nancy's philosophy is steering away from insofar as it entails a sense of closure in thought and, by extension, in possibility, and a simplistic and artificial opposition between thinking and doing, between philosophy and action and between the abstract and the concrete – oppositions that are in direct conflict with his ethico-political thinking and inconsistent with the open-ended nature of being(-with).[15] So, according to Nancy, the question 'what is to be done?' is a sign that one no longer stops to ask 'what is to be thought?' as though 'one already knows what it is right to think, and that the only issue is how one might then proceed to act' (Nancy, 1997: 157). In this case, 'what is to be done?' becomes merely a question of calculating, programming and applying pre-given rules and theories in order to achieve pre-set goals and objectives or pre-empt future events and occurrences. By implication, 'transforming the world' becomes a matter of realising an already given interpretation of the world. But as Nancy puts it,

> Perhaps we no longer even know what it is to think nor, consequently, what it is to think 'doing', nor what 'doing' is, absolutely. Perhaps though, we know one thing at least: 'What is to be done?' means for us: how to make a world for which all is not already done (played out, finished, enshrined in a destiny), nor still entirely to do (in the future for always future tomorrows).
>
> (ibid.)

Not already done and not still entirely to do: what we are left with, then, is a world-in-the-present, a world to be invented and created 'immediately, here and now, at every moment, without reference to yesterday or tomorrow' (ibid.: 158). This is indeed the essense of Nancy's philosophy. It is a call for being present *to* the present; freed from the shackles of the past; and totally open to the uncertainty of the future.[16] It is a call 'summoning us to an openness without anticipation, a preparedness for surprise that could never eradicate surprise, a world in which incertitude and undecidability are understood to be definitive of the human condition' (Hutchens, 2005: 160). From such an openness erupts a *sense* of freedom that is neither a property nor a political condition, but 'a being, an existentiality' (Nancy, 2005: 164): freedom that 'produces itself as existence in accordance with relation' (Nancy, 1993: 78), that is, in accordance with the spontaneous *clinamen* engendered through and within the experience of being-in-common. It is an *event* of existence that *happens* and *takes by surprise*:

> when I am acting, doing something, there is always something new, unexpected, in what I do and/or in myself as the agent: the text I write, the words I say, the face I show to my partners, etcetera are always surprising to myself. I cannot say 'I did not intend to do this or that'; I must take it as 'mine' because it did occur to 'me' and 'I' am made only by such occurrences. Then, this 'surprising' freedom is not libertarian freedom, which presupposes a free subject before any acting. One could say: to be free is first of all to be free of the self.
>
> (Nancy, 2005: 164)

This freedom is, in fact, freedom *from* (neo)liberal freedom, or more generally, freedom from 'established freedoms': 'a freedom that would not guarantee political, religious, and other freedoms, but an *inaugural liberation* with respect to these freedoms, insofar as they would be nothing other than the freedoms of choice at the interior of a closed and pre-constrained space' (Nancy, 1993: 79). Nancy's rethinking of freedom in terms of surprise and existence (rather than as a property of the subject that is guaranteed through establishments, safeguarded by frameworks and securitised with technologies) remains very much in tune with his rethinking of the political as the *opening* of a space. This space is not determined in advance but opened *singularly* and *each time* by freedom itself, freedom as an initial, inaugural and arising event of existence – freedom as 'an initiality of being' (ibid.: 78). Correlatively, the task of politics, according to Nancy, does not lie in providing structures and principles for this freedom (as this would subject it to yet another establishment), but in liberating it from every establishment and (re)opening the space of its inaugural sharing (ibid.: 79). To do so, politics has to re-conceive freedom in terms of 'uncertainty' as a condition that is carved in the heart of human existence itself (Hutchens, 2005: 157), and re-imagine the future as a space that is 'wholly beyond the reach of free agency [and] resulting from incessant surprisings of experience' (ibid.). It also has to aim at the *communication* and *interaction* between singular beings rather than at the *identities* of subjects and citizens.

So, how do all these reformulations relate to the overall context of this book and what do they add to our previous understandings? Well, as Hutchens (2005: 8) puts it, 'Nancy gives us nothing that can be applied elsewhere, only the precious gift of an alternative perspective'. In my opinion, this is precisely the gift that current policies and governmental strategies are in desperate, though unavowed, need of. This is also the gift that both governmentality studies and Agamben-inspired approaches, to the issues of asylum, immigration, citizenship and biometric identification, can benefit from. For, it offers a space of *contemplation* where one can pause to reflect *intuitively* rather merely logically upon the nuances and depths of these issues and to dive into the root causes of violence rather than merely surf on its capricious and mounting waves. One can, indeed, readily see how Nancy's *meditations* on community and the political provide an opportunity of *re-framing* what is cast as a 'problem' and a 'solution' within contemporary forms of governance, and thereby offering ways of transforming that 'mental schema' of politics we mentioned in an earlier chapter.

More specifically, Nancy's politics of openness and surprise can serve as a potent antidote to the prevailing politics of fear and control. As discussed throughout this thesis, the vacillating borders between the inside and the outside, the legitimate and the illegitimate, the included and the excluded and so on, are often mediated through and constructed according to an emotional landscape of fear, distrust and suspicion. This landscape induces a governmental desire for mastering the world and the future which in turn leads to the 'fictionalisation of the world' (Bigo, 2006a: 58) whereby everyone ends up being dragged into a space of simulacra and projections in which paranoid scenarios and apocalyptic scripts loom large. What follows is the mobilisation of an array of pre-emptive techniques and securitisation technologies, and the construction of various images of otherness and dangerousness, all being based upon the immanentist belief that one can create 'a grammar of *futur antérieur*' by which the future can be read as a form of the past in order to manage risk and prevent unwanted events (ibid.: 61). While such attitudes towards the world and the future are operating in the name of security, safety and the fight against terrorism, they are also lacking a sense of *awareness* that 'the present situation is also the fault of the will to master the world, to try to control some part of it by "scientifically" discriminating the enemy within, and to believe that technology can do it' (ibid.: 62). They are a manifestation of the closed immanentist style of thought, which stifles openness to the unknown and instils fear of difference and otherness.

Against such ways of approaching the world and the future, Nancy maintains that '[w]hat will become of our world is something we cannot know, and we can no longer believe in being able to predict or command it' (Nancy, 1997: 158). This, however, does not amount to resorting to a *passive passivity* (we all know the age-old trick: bury one's head in the sand), but to a kind of *active passivity*, a passivity in terms of surrendering that desire to control the future at any cost (including the cost of life itself) and in welcoming the new and the unknown, and it is active in the sense that one 'can act in such a way that this world is a world able to open itself up to its own uncertainty as such' (ibid.). This, in fact, places an ethico-political imperative 'to invent a world, instead of being subjected to one, or dreaming of another. Invention is always without model and without warranty. But indeed that implies facing up to turmoil, anxiety, even disarray. Where certainties come apart, there too gathers the strength that no certainty can match' (ibid.). This is a strength that stems from vulnerability itself rather than power,[17] from surrender rather than control, from the burst of freedom understood as the initiality of being (an invention) rather than a property or a right and from coming to terms with the fragile, finite and uncertain nature of our being-in-the-world rather than resisting or avoiding it. Technologies of securitisation (including those of biometrics), prudentialist mechanisms of control, measures of *anti-* and *counter-*terrorism, are all *suppressive* in nature. In their quest to pre-empt perceived threats and *cure* the ills of society, they merely function by way of suppressing surface symptoms (and often with violent side effects) rather than *healing* the deep layers underlying the present situation. One cannot suppress, control and manipulate the world without doing damage to the openness and spontaneity of the world.

Reframing our relationship with the world and the future in terms of surprise, openness and uncertainty has the potential to alleviate some of the fears and anxieties, which continue to feed the cycle of violence and suffering and justify the increasing technocratisation of the political, and to provide a means for imagining what Welch and Panelli (2007: 352) refer to as 'new geographies of singularity and collectives'. Viewed through the lens of Nancy's philosophy, such geographies would not be predicated on the politics of borders, the logic of securitisation and the myth of common being, but on generous, open, indeterministic and non-assimilationist modes of relating and communicating. They would allow the *free* circulation of singularity and difference within the dynamic fabric and energetic space of being-in-common, and counter forms of marginalisation and prejudice that are associated with 'Othered cultures' (ibid.: 353). As should be clear by now, in the Nancyan vocabulary, there is no 'us and them' let alone 'us versus them'. There is, instead, only 'us' as a plurality of singular beings, us *with* us (being-with), us *among* us (singular plural) and us *toward* us (*clinamen*). There is, in this formulation, 'comfort for each being because it suggests an ultimate [...] inclusivity. But it is also a challenge, for it exposes and emphasises the diversity and existential qualities of singularity (i.e. not only the plurality of singularity but also the distance and difference between beings)' (ibid.). In this sense, the challenging task for politics is very much a matter of seeking to combine two aspects which, as Schwarzmantel (2007: 472) puts it, 'pull against each other', that is, affirming an idea of equality among all beings, and at the same time, encouraging a community of difference instead of a community of unity and common being, since the latter, as discussed earlier, threatens diversity by enforcing modes of homogeneity, assimilation and conformity and by endorsing certain identities while depicting others as dangerous.

In considering the dialectic of universalism and particularism, Dallmayr (1997: 189) invokes a similar argument, which he sees as expressing 'the need for a differential strategy beyond global synthesis and its denial: that is, for a kind of double move or double gesture proceeding cautiously in the interstices of affirmation and negation'. So it is not simply a matter of *universalising* the rights and equalities of the majoritarian culture, for, after all, universalism, as Waltterstein (in ibid.: 190) notes, is ' "a gift of the powerful to the weak" which confronts the latter with "a double bind: to refuse the gift is to lose; to accept the gift is to lose" '. Nor is it simply a matter of preserving and reinforcing a sense of particularism as this often leads to self-enclosure and the ghettoisation of communities. The challenge is instead a matter of intervening at the *symbolic, imaginary* and *representational* levels rather than merely at the level of political practices in order to create a *mental shift* in the way in which the other is perceived and responded to. It is a matter of heightening awareness that 'if we are already intimately involved with all others to the extent that they are not others but part of "us", then it is a thinking (and a politics) that conveys and (re)produces an impression of separateness that leads to our difficulty' (Edkins, 2005: 382). This also entails heightening awareness of the fact that constructions of community that are based on the idealised notions of agreement, unity and conformity are not only impossible but also dangerous, and that relying on technology as an ultimate solution only ends up glossing over the structural and deeply rooted socio-political problems, including those of inequality, ostracisation, alienation and violence – all too often, technology merely functions as an 'improved means for unimproved ends' (Webster, 2000: 86) and too much planning, organisation and management only yield 'increased "rationalization" both in Weber's sense and Freud's sense – with the outcome pointing to a worst-case scenario: "The Iron Cage *and* self-deception" ' (Dallmayr, 1997: 192).

In practical terms, and without instrumentalising Nancy's thoughts, handling such challenges would involve working closely with policy makers, advisers and analysts to instil an appreciation of the ontological position of singularity, increasing their understanding of how it unfolds within and links to everyday experiences, practices and encounters, and encouraging them to 'support populations to grasp the principle that being-in-common is the only collective state that can be realised, and to explore with them the ways in which collectives can be established that *directly and effectively reflect* this reality' (Welch and Panelli, 2007: 353). It would involve 'a fundamental re-evaluation of what it means "to be" – in singular-being and collective forms' (ibid.). It would involve the laying bare and the challenging of subconscious assumptions and myths that continue to inform present policies vis-à-vis otherness, threat, security and the conceptualisation of community on the basis of common being. It would involve countering the economy of fear and enmity, currently pervading political discourses and practices, with the possibility of solicitude and compassion, and a new imaginary of being-with that fosters an ethics of engagement, communication and generosity whereby difference, disagreement and disunity are not seen as a threat but an integral part of the dynamics and intensities of being-in-common. Echoing Nancy's 'counter-managerial vista', as Dallmayr (1997: 192) puts it, Wallerstein argues that perhaps

> we should deconstruct [systems] without the erection of structures to deconstruct, which turn out to be structures to continue the old in the guise of the new. Perhaps we should have movements that mobilize and experiment but not movements that seek to operate within

the power structures of a world-system they are trying to undo. Perhaps we should tiptoe into an uncertain future, trying merely to remember in which directions we are going.

(in ibid.)

Actualising such imperatives is certainly an enormous and difficult task but also a necessary one if today's societies are to meet the challenges of a world in which democracy is continuously being eroded by the techno-economic dominion and where the political is increasingly being reduced to the play of neoliberal forces and practices. It is a task that requires a responsible, accountable, patient and passionate (rather than merely rational and calculating) engagement with the world, and a careful and mindful examination of how our (in)actions and interactions affect the material fabric of our being-with. As Nancy (1991: xli) alarmingly argues,

> if we do not face up to such questions, the political will soon desert us completely, if it has not already done so. It will abandon us to political and technological economies, if it has not already done so. And this will be the end of our communities, if this has not yet come about. Being-*in*-common will nonetheless never cease to resist, but its resistance will belong decidedly to another world entirely. Our world, as far as politics is concerned, will be a desert, and we will wither away without a tomb – which is to say, without community, deprived of our finite existence.

We certainly deserve at least a better *end*!

Conclusion

Throughout this chapter, and after establishing the importance of the notion of community to the rethinking of citizenship, we have embarked upon an exploration of a different vision of community, one that goes beyond the endless vacillation between communitarianism and liberalism. This vision is based on Jean-Luc Nancy's work and finds in the ontological dimension of relationality and singularity the condition of ethics and the essence of the political. By drawing on concepts such as being-with, being-in-common and being singular plural, Nancy offers a co-existential analytic in which community is understood as neither the gathering around a common substance nor the organisation of sameness, but as an ongoing and dynamic process of sharing, communicating and being exposed to one another's singularity, vulnerability and finitude. It follows from this that the political, far from being a matter of creating operational systems and maintaining organisational structures for the management of society, is primarily a question of relation and an opening of the space between singular beings. Re-imagining community and the political in this way offers an alternative view of a 'political community' that is not circumscribed by the fixtures of a national identity or bound by the demarcations of territorial and economic boundaries, but one that can be construed as a space of mutual exposition, dynamic engagement and a spontaneous sharing of singularity and difference.

Such reconceptualisations, we argued, carry the potential to liberate politics from the pervasive fears and escalating anxieties that weigh heavily upon it, close off its horizon and reduce its function to the mere implementation of securitisation strategies and pre-emptive mechanisms. They promise new geographies and collectivities that are based on openness, generosity and respect for alterity, and which may as well cancel the need for borders and identification technologies, and reinstate the question of relation and sociation at the heart of politics. For *how* we relate to and engage with one another is the political question *par excellence*.

Conclusion

This book has looked at the example of biometric identity systems and other related developments in order to explore and elucidate some of the transformations pertaining to the field of governing. In doing so, it has drawn attention to the specific practices, techniques, processes, rationalities and mechanisms by which central issues, such as identity, citizenship, access, entitlement and the like, are currently being managed, organised and controlled. As an overall framework of analysis, the book has engaged with the notion of biopolitics as an umbrella concept for thinking about the characteristics, challenges and stakes of managing the individual-body and the population-body through technology and in the name of risk and security. Rather than being faithful to one specific version of biopolitics, the book has attempted instead to bring together differing interpretations of and approaches to the notion of biopolitics so as to account for the multifaceted and, at times, paradoxical nature of contemporary forms of governing the life of the individual and the population. This paradoxical dimension is mostly apparent when the figure of the citizen is starkly juxtaposed with that of the asylum seeker (and other related figures such as the refugee and the immigrant). Such juxtaposition has thus been taken up throughout the entire journey of this research.

One of the issues addressed in this book has to do with the function creep of biometrics and identity systems. Through the Agambenian take on biopolitics in terms of the notion of exception, we began by suggesting that current policies of immigration control and border management demonstrate this logic of exception whereby *decisionist* interventions and illiberal practices, such as biometric profiling and fingerprinting of asylum seekers and their incarceration in detention centres, are being enforced upon certain groups and in ways that render their exclusion from the official juridico-political structure as an 'inclusive exclusion'. Importantly, we argued, this rationality of exception is not tantamount to the casting out of politics or the cancellation of what is deemed as the norm, but represents an integral part of 'normal politics' itself and a condition of possibility for its actualisation and organisation. It is also not only a matter of confinement but that of 'knowledge production' as well, which provides the basis and means for control strategies and pre-emption mechanisms such as those performed through asylum smart ID cards (ARCs).

The function creep of biometric technology and identity systems has, therefore, been approached in terms of their spillover from exceptional domains and into the general population-body through the introduction of national identity cards and the proliferation of biometric procedures. Yet, as we deduced from the amalgamation of different empirical examples, this spillover is not to be understood as the turning of exception into the rule in a somewhat homogenous way, but more as a polysemic process, which involves directing selective and refined modes of surveillance and control at certain groups while normalising the rest of the population. In this, we identified a number of paradoxes and multi-layered functions, notably in the way in which liberal and illiberal practices tend to concurrently conjoin. Bigo's notion of the banopticon has been deployed as an analytical tool for attending to such nuances and variations. Through this *dispositif,* we have also been able to merge oppositional approaches to biopolitics without losing their specificity and distinct insights.

A book on the theme of biometric identity systems cannot do without a thorough discussion on the theme of identity itself. So, the second study has been devoted to exploring precisely the question of identity in relation to biometric technology and with a particular focus on the field of asylum management. Starting off the discussion with the oft-posed question 'who are you?', this section of the thesis went on to examine the kind of identity biometrics is concerned with. Through an engagement with a cluster of philosophical concepts and postulations (particularly Cavarero's and Ricoeur's respective differentiation between the 'what' and the 'who', the 'idem' and 'ipse' dimensions of identity), we drew a distinction between the kind of unique identity that biometric technology claims to protect, where the person is reliably identified as the same over time and in a variety of different situations, and the more intricate sense of identity where uniqueness refers to the story one tells about oneself in the presence of others. Our examination has led us to conclude that although, at first sight, biometrics may seem to be mainly concerned with the 'what' and the 'idem' aspects of identity, the dimensions of 'who' and 'ipse' are also implicated in a mutually transformative way, such that these differing poles of identity cannot be simply opposed and sharply demarcated.

Through this examination, we have also identified some of the bioethical challenges emerging out of the deployment of biometric techniques for the management of asylum. We cast these challenges in terms of the ethical implications concerning biometrics' attempt to extract singularity from body parts and to bypass the self-attesting story. We argued that such a reductionist approach towards identity and uniqueness leads to the occlusion of the narrative dimension of identity and with it the amputation of the possibility of solicitude and sympathy in decision-making. As a response, we proposed a bioethical approach based on the narrativity thesis as both a critique of biometrics' foreshortening of the narratable aspect of the person, and a call for a more compassionate and embodied engagement with the uniqueness of each refugee story within the field of asylum policy and management. We also addressed some of the limitations of this approach, specifically with regard to the constraining relationship between narrative and norms (be they linguistic or social), and the broader institutional and structural context that mediates between the asylum seeker and the immigration officer, and frames their power dynamic.

To explore further the multifaceted features of biometric technology and ID cards, Chapter 4 examined other problem fields in which biometric technology is activated as a solution. Moving from, but without leaving behind, our discussion on the 'exceptional' space of asylum and immigration, this chapter went on to tackle some more routine sites and practices whereby identity is securitised through biometric technology, this, by shifting the focus towards the figure of the 'citizen' and considering what is entailed in the recasting of the citizen's identity in terms of risk and security. One of the issues explored in this chapter relates to the phenomenon of identity theft and fraud. Drawing from the governmentality perspective and through an examination of relevant policy documents and current identity protection mechanisms, we argued that the fight against the threat of identity fraud and theft is partaking of a wider strategy of risk management and is informed by specific governmental rationalities whereby individuals are increasingly being made responsible for managing the risk of 'identity crime'. Identity, in this context, is viewed as a 'commodity' that is integral to one's ability to carry day-to-day activities and whose management and protection relies on the know-how of the individual. Cumulatively, biometrics and identity systems are being promoted as an important element of the knowledge toolkit involved in the fight against identity fraud and theft. They are formulated as a technology of subjectivity that allows the individual to proactively securitise her identity and exercise her freedom of consumption within the market circuits. Such a rearticulation of identity in terms of risk, security and technology has a bearing on the way in which citizenship itself is understood and practiced. The remainder of Chapter 4 has, therefore, been focused on exploring what kind of citizenship is the 'biometric citizenship'.

Through the example of the UK IRIS system, we began by suggesting that biometric citizenship is a neoliberal citizenship to the extent that it is embedded within the principles of choice, freedom and active entrepreneurialism that are hallmarks of neoliberalism and its market-driven ethos. The voluntary use of advanced border technology, such as IRIS, by mobile neoliberal actors illustrates well the functioning of this type of citizenship in that it allows the flexible and speedy exercise of rights (for instance, the right of freedom *qua* mobility) at the border beyond the traditional territorially based entitlements and through the deployment of biometric technology. Here again, instead of assuming the homogeneity of this paradigm, we stressed upon the multidimensional and aporetic nature of biometric citizenship as a neoliberal citizenship, this, by reconsidering the intimate interplay between exception and neoliberalism, and the constitutive relationship between the figure of the citizen and that of the asylum seeker, illegal immigrant and so on.

In addition to its neoliberal aspect, biometric citizenship is also a biological citizenship. This is not only in terms of the use of the body itself for identification, but also in the way in which biometric technology is deployed as a means of sorting through different forms of life according to their degree of utility and legitimacy in relation to market economy, and ultimately distinguishing between those to be included and those to be excluded. However, the biological aspect of biometric citizenship is not reducible to this bio-economic factor. The racial and national dimension is equally relevant. In terms of race, we argued, following from Pugliese, that the very technical infrastructure of biometric technology is embedded within race-centred practices and operating, at times, according to a quasi-universal template of whiteness. This is evident in some of the scientific literature in which some bodies are presented as being inscrutable and illegible to the biometric scanning machine (as in the case with 'failure-to-enrol', for instance). Pre-emptive profiling mechanisms, performed through biometric technology, also rely on racial categories and carry with them the risk of discrimination. As regards the national dimension, this often takes the form of a symbolic capital, rendering biometric identity cards, for instance, as a token of national belonging, and in some contexts, a means of singling out specific groups as part of a double movement of inner exclusion and outer inclusion.

Much of the dynamics that underpin biometric citizenship is driven by fear and distrust towards an array of 'dangerous others'. Affects, as such, play a prominent role in governing, and constitute another important feature of citizenship. Building upon the work of Isin (2002), we argued that biometric citizenship is also a neurotic citizenship. At one level, and as with the example of border control, neurosis seems to be emerging as an object of analysis and scrutiny within risk management techniques. Developments such as those of biometric behavioural profiling feed upon an imagination of a neurotic subject whose moods and emotions can be measured and quantified in order to detect suspicious anxieties and hostile intents. They are not only concerned with capturing the body's surface for the purpose of identification, but also with probing into one's interiority and (un)conscious thoughts so as to anticipate behaviour and prevent action in advance.

At another level, the neurotic aspect of biometric citizenship can also be witnessed in the discourses and practices surrounding identity fraud prevention. Here, affective accounts of those who have fallen victim to identity crime are often invoked in ways that heighten anxiety about this issue and thereby encourage citizens to be more vigilant, responsible and proactive in protecting their identities. To this end, biometric identity systems are promoted as a prudentialist solution to identity fraud and theft. They are invested with emotional qualities, acquiring the status of a tranquilising mechanism and with it more legitimacy and significance.

Such transmutations in the ideal and practice of citizenship through biometric technology are indicative of the thinning process by which citizenship has become increasingly a question of implementing technical mechanisms of identity management rather than a matter of active participation in the political sphere. They thereby reveal the symptoms of a growing technocratisation of politics and the weakening of the ties between the ideal of citizenship and that of political community. Using the governmentality thesis as an analytical tool has helped us unpack the discourses, practices and rationalities pertaining to such symptoms. Yet, it remains that the task is not simply a matter of exposing these symptoms and looking at their concrete manifestation in different sites of governing. Nor is it simply a matter of regarding citizenship or community as a *project tout court*. One must go further. A deeper reflection on and a more deconstructive approach to what the idea of citizenship and community *means* in an age of erosion of nation-states and the advent of biotechnological operations is also an important and necessary task. For this enables the laying bare of the self-evident and deeply embedded assumptions and beliefs that support the perpetuation of certain illiberal, exclusionary and violent practices in the name of citizenship/community itself, and which continue to feed into the fear-mongering measures of securitisation. This also enables the placing of the debate on biometric identity systems in a broader context that goes beyond, but without excluding, the governmentality perspective and its lines of reasoning. Such was the task of Chapter 5.

So instead of being content with a surface analysis of the discourses, practices and rationalities concerning biometric identity systems, we have sought, through Chapter 5, to rethink some very fundamental issues and deeply rooted categories that are at the very base of Western politics and its attitude towards otherness. By the same token, we have also sought to bring out some wider and profound ethical and political dimensions that are pertinent, though not in an immediately obvious manner, to the rationale and motives behind the increasing fascination with biometrics and identity systems as a security solution. Putting the notion of community into question, and from an ontological standpoint, has provided the touchstone for this task.

Drawing upon the work of Jean-Luc Nancy, we began by demonstrating how current policies of immigration and citizenship epitomise the logic of what Nancy terms immanentism, that is, a conception of communality based on self-enclosure and self-identification. We argued that the manifestation of this logic, in the field of immigration and citizenship, unfolds through the act of bordering, technicism and the invocation of mythical collective identity, and is always presented as a *work* to be accomplished. In many respects, immanentism does echo some aspects of the totalitarian side of biopolitics. Against the immanentist modality, Nancy proposes alternative ways of reconceiving community that are neither reducible to the formulations of communitarianism nor to those of liberalism. The rest of Chapter 5 has, therefore, been devoted to exploring a cluster of ontological concepts such as being-with, being-in-common and being singular plural through which Nancy seeks to emphasise the important role of relationality, sharing and finitude for rethinking the essence of the political and the question of community.

Nancy's ontological stance vis-à-vis the political also carries a normative dimension whose premise is primarily based on an anti-managerial attitude towards the relations between singular beings, an interruption of that which is deemed as workable and practical, as well as a demand to remain open to the uncertainty and unknownability of the future. So rather than considering community as an operative structure or an organisation of sameness, Nancy provides a vision of community as an ongoing process of sharing and communication between singularities. Rather than reducing the political to the socio-technical forces and operations of politics, he contends that the political is mainly a question of relation, an opening of the space between singularities. Rather than regarding the future as being amenable to control and pre-emption, he insists that one must respect and embrace the element of surprise, the burst of the unknown and the irreducibility of the uncertain that are part and parcel of the open and spontaneous character of the future.

Reframing community, the political and the future in this way offers us a different onto-normative vocabulary for approaching the ethical stakes of current technological developments and contemporary policies of immigration and citizenship. More specifically, this reframing allows us to see, even more clearly, how biometric identity systems are but technical expressions of a politics of fear, distrust and suspicion that close off the horizon of the future through their pre-emptive and calculative techniques, and gloss over prevailing inequalities and the lack of accountability and awareness towards the root causes of the ongoing socio-political problems. In this sense, we argued that Nancy's approach could act as an antidote to such a politics by stressing the importance of relationality over technicality, of communication over fusion under one body, of singular beings over prudential subjects, of openness over fear and closure and of responsibly engaging with the deep layers, rather than merely with the surface, of current challenges and difficulties. If viewed from the vantage point of the Nancyan political ontology, biometric identity systems would not only seem as redundant in resolving the problems facing contemporary societies, but also as an obstacle to tapping into the strength entailed in being able to open up to the uncertain and the unknown, and into 'the capacity to engage with material and social worlds in imaginative and generous, rather than fearful, ways' (Diprose et al., 2008: 285). Politics, in this sense and instead of investing most of its effort on enhancing pre-emptive technologies of securitisation to keep the perceived dangerous other at bay, needs to direct its endeavour to developing 'new sensibilities toward otherness [...] that may well end a fundamental ontological difference between [the West and its others] without at the same time reducing various cultural "zones" to an equally fundamental sameness' (Isin, 2002: 15). It needs to aim at facilitating the dynamic interactions between singular beings by creating spaces for open dialogue, for listening, sharing and being-with in ways that do not favour unity, consensus and conformity but encourage an acceptance of and respect for disagreement, heterogeneity, alterity, dissonance and so on. Developing richer ethical ontologies the way Nancy suggests, and as Mills (2010) puts it, will certainly help such an undertaking and enable a better understanding of the practical exigencies of everyday life and more responsible and innovative responses to its normative demands than what technical solutions can afford.

Such an endeavour demands that we courageously venture beyond the strictly defined contours of traditional bioethics in order to rework its fundamental concepts and explore other ways of approaching, thinking and doing ethics. As I mentioned in this book, the majority of debates on the bioethical implications of biometric identity systems have been couched in the familiar terminology of privacy, rights, liberty and autonomy, notions that are founded on the model of the liberal individual and central to the utilitarian approach of moral philosophy. But in drawing exclusively on such concepts, these debates often fail to recognise other important elements whose ethical dimensions are so relevant to the challenges raised by contemporary technological developments. Engaging with the field of Continental philosophy is one way of addressing this deficit insofar as it provides a means of retrieving 'the often-neglected political and historical background to some of the key debates in bioethics today' (Mills, 2010), reshaping the debate on what counts as a bioethical issue and developing a new grammar of the normative, 'one that revolves around notions of vulnerability, interdependence, embodiment, singularity, forms of life and biopower' (ibid.). Such was the grammar adopted in this research.

Finally, I hope that I have managed, through this book, to expose the complexities and diverse paradoxes inherent in contemporary modes of governing through technology. As I demonstrated, advances in security technologies such as those of biometric identity systems cannot be understood from a merely one-dimensional perspective or as belonging to one modality of rule. They are embedded in multifaceted, context-dependent, interrelated and competing paradigms of control that are operating in often unjust and unequal social and political circumstances. So, in part, the book makes a contribution to the sociological study of security and to the debates on biometric identity systems by acting as a corrective to some of the totalising and universalising theories that tend to overlook the manifold contextual nuances and dilute singular cases into the metanarrative of surveillance and the like. This book also makes a conceptual contribution to existing research on biopolitics, bioethics and biotechnology by providing interesting signposts for navigating through these fields and by bringing together different philosophical approaches, conceptual tools and empirical material into dialogue with each other. In doing so, the book demonstrates, I hope, the value of methodological and conceptual experimentation for critically engaging with the questions and concerns emerging from this area of research and for bridging the 'perceived' gap between the philosophical and the empirical, the abstract and the concrete. Viewed more broadly, the book can also be seen as contributing to theories of identity by offering a thorough consideration of issues revolving around embodiment, subjectivity and singularity, and revealing the multi-layered and polysemic aspects and functions of identity in its interplay with governance and technology.

Of course, such a contribution does not exhaust the different ways by which the theme of biometric identity systems can be approached and researched, nor does it answer all the questions pertaining to this field. If anything, this book is leaving much to be asked if only by way of inciting interest in exploring and developing new alternative approaches, conceptually as well as empirically, for rethinking the political and ethical stakes of security mechanisms and identification technologies, and exploring their manifestation from a variety of angles. One line of enquiry, for instance, that is worth pursuing further is to do with the sites and practices of resistance (rather than merely those of governing) that are emerging as transgressive responses to the stringent policies of immigration and citizenship and the increasing deployment of biometric technology for border security. A detailed phenomenological study of the way in which different groups of the population make sense of and experience their embodied biometric identity also warrants further investigation.

On a personal note and reflecting back on the journey I have undertaken throughout this research, I can see how this has been a *transformative* experience, one that involved many torsions and ruptures. Engaging with the issues addressed in this book in an embodied and empathic way has put my own sense of identity into question and shaken the very foundations of the 'who' I (thought I) was. More so, this experience has put me beside myself and taken me outside of myself in ways that seem, at the moment, rather irreversible and profound. For, once the refugee within each of us has been encountered and touched, once her uncanny presence has been acknowledged and felt, she can never disappear – as she finds a permanent inner place of dwelling in the midst of worldly homelessness. Once the neoliberal subject has been confronted with the limit, its anxieties and certainties, its fears and dreams can never have the same hold again. And, once the singular being has been exposed and abandoned to the *sense* of existence, she becomes more *aware* than ever, more *receptive* than ever of the intensities and fragilities, of the joys and pains of being-in-the-world-with-others.

Notes

Introduction

1. For example, the Jews in Nazi Germany, blacks in South Africa under apartheid, Palestinians in Israel and North Africans in France. Elsewhere, Lyon (2004: 2) points out that '[a]t the extreme end of the spectrum, ethnic classification on ID cards was directly connected with twentieth-century genocide in several countries, including Rwanda'. There is a need, however, to keep in mind the contextual differences regarding the use and the objectives of these various identity projects.
2. Although in the context of security, the digital body has always to be mapped onto a vital body since it is the latter that is capable of the actions that security mechanisms attempt to pre-empt.
3. To which bodily integrity applies in cases of physical intervention or procurement of test materials.
4. For example, biometric templates, genetic profiles, digital images of fingerprints and so on.
5. See, especially, the work of cyberfeminists such as Haraway.
6. For Schmitt, 'the political' refers to the specificity of politics. Some French theorists and philosophers such as Nancy and Lacoue-Labarthe differentiate semantically between politics and the political through the use of the articles *le* and *la* (*la politique/le politique*). I shall return to this in Chapter 5.
7. Of course, this is only one way of understanding 'the political'. The division between friend and enemy does not exhaust other meanings and functions that the political may assume. For instance, in Chapter 5, we shall explore a different take on the notion of the political based on Nancy's conceptualisations where the essence of the political is not so much a matter of performing a division between friend and enemy, but a question of relation and being, a socio-ontology, so to say. From my part and in the context of later discussions, I prefer to subscribe to the latter framing given its ethical dimensions, to which I shall return in due course.
8. Again, Nancy's perspective on the space of the political problematises the neat demarcations between 'the social' and 'the political'.
9. See John Schwarzmantel (2003).

1 Biometrics: The Remediation of Measure

1. By relying on the 'vital' body for identification.
2. For example, digital photography refashions traditional photography as well as its different uses.
3. Akin to that which Thacker attempts to achieve through the notion of biomedia.

4. This, however, should not undermine the important role of 'names' in the history of identification. For a critical discussion on the relationship between names, identity and technologies of writing, see Derrida (1976).
5. See also Kaluszynski (2001), Rabinow (1996) and Gates (2005).
6. See Chapter 2.
7. After almost 10 years since their first proposal, the UK biometric ID cards scheme has been scrapped in early 2011 under the Coalition Government. For a detailed discussion on the different reactions and arguments vis-à-vis this scheme, see Wills (2006). Here, Wills argues that the campaigning and oppositional discourses, which were directed at biometric ID cards in the UK, tended to cluster around three different strands of concern: pragmatic and technical (e.g. NO2ID); principled (e.g. Defy-ID and Privacy International); and financial (e.g. the LSE Identity Project).
8. With the exception of members of military forces, prisoners and the mentally ill – these groups were already listed on specialised registers.
9. Elsewhere, Foucault (2003 [1976]: 241) suggests that the sovereign's old right of taking life or letting live was not *replaced*. Rather, it came to be *complemented* by the new right of making live and letting die. Foucault's ambivalence vis-à-vis this point has crucial implications with regard to subsequent debates on the concept of biopolitics and its various theorisations and interpretations.
10. Foucault's examples include socialism and Nazism. But a more contemporary example can be found in immigration policies, which hinder the circulation of people from poor or developing countries, exposing some to life-threatening experiences (for instance, the death of 'clandestine' migrants trying to cross the borders of Europe and America).
11. See also Catherine Mills (2004) and Michael Dillon (2005).
12. Here, one can see a certain similarity emerging between Agamben's statement and Foucault's above-mentioned articulation of biopower in terms of life and death. For while each of them is resorting to a different strategy to understand the phenomena at hand (Foucault through racism and Agamben through bare life), they are both reaching the non-concluding conclusion about the 'aporia' of biopolitics (same techniques that are designed to preserve life can also be used to destroy it). Yet, for Foucault, this aporia remains quite a puzzle and he does not propose any means by which it can be understood. Agamben, on the other hand, suggests the logic of the state of exception as a possible explanation. For this reason, Agamben regards his thesis as an attempt to 'correct or at least complete' the Foucauldian one.
13. Some, on the other hand, see in this a flattening of Foucault's materialist genealogy of power and its 'contextual' and 'historical' dimension (see, for instance, Rose and Rabinow, 2003).
14. Again, I place emphasis on immigration, asylum, detention, borders and so on.
15. Such as the death of the Iraqi civilians in the American attacks on Haditha, Fallujah, Najaf and so on; the death of those who tried crossing the deadly borders of the 'West'; the deaths inside camps and detention centres (Sangatte and Harmondsworth, for instance); and (sadly) the list goes on.... Of course, these deaths are not the same, nor can they

indistinguishably be squeezed into the figure of the *homo sacer*. But what they all share in common is their unpunishable feature.

16. Unlike the case in Agamben's account.
17. See also Rose and Novas (2000).
18. A term coined by Catherine Waldby (2002). It refers to the process by which life itself is rendered as a productive and profitable economic and political value through the enhancement of health.
19. See also Rabinow's notion of biosociality.
20. Or, perhaps, on 'specific domains' in the life sciences. For life science is now a broad field encompassing, as Dillon (2005: 41–2) suggests, 'molecularised biology, through digitisation[,] social and managerial sciences of development now prominent in the fields of global governmentality, global development policies, human security and even military strategic discourse'.

2   Homo Carded: Exception and Identity Systems

1. Relating to the issue of data sharing and the ability to enrol a user with one vendor's technology and verify the same user with other vendor's technology (see http://www.bioscrypt.com).
2. It must be noted here that my choice of these examples is not merely a 'practical' one, but also stems from a 'political' concern with that which emerges out of the interface between biometric technology and the immigrant/asylum body. It follows Agamben's suggestion that in order to understand the functioning of modern (bio)politics and its various technologies, one should start not from the figure of the 'citizen', but from those liminal figures whom, for one reason or another, are excluded from the official political order and yet remain attached to it. Framing the question as such will allow me to expose some of the forms of violence that are inextricably linked to such policies without, however, obscuring the complexity and the contradictory character of their rationalities.
3. Although it is argued that 'if the Dublin Convention puts an end to "refugees in orbit" in the European Union, Member States still contribute to this phenomenon in the rest of the world' by sending an applicant for asylum to a host third country (Hurwitz, 1999: 650).
4. Worth noting here is that the European Parliament has initially delineated very strict limits to the use of the Eurodac system: 'the results of the fingerprint comparisons transmitted by the C.U. [Central Unit] are lawfully used exclusively to ascertain the competence of the Member State of origin in accordance with the Dublin Convention' (cited in van der Ploeg, 1999a: 299).
5. Although now this *forensic* aspect of fingerprinting has lost much of its exclusivity and specificity as the practice of fingerprinting is increasingly being adopted for a wide range of uses (van der Ploeg, 2003a: 60).
6. Worth stressing here is that such examples demonstrate that these trends have a track record that dates back to the 1990s and they are, therefore, not only a response to the events of September 11 or a materialisation of the rationality of the so-called 'war on terror' (although the latter have undeniably contributed to the intensification of these trends).

7. However, and as we shall see, in addition to this dual regime of circulation, there are further divisions, further segmentations based on the principle of 'economic triage', which operates through various migration programmes such as 'Highly Skilled Migrant Programme', 'Sectors Based Scheme' and other points systems. These programmes complexify the simple duality of inclusion and inclusion.

8. A term borrowed from his mentor Louis Althusser.

9. And here, it depends also on what African passport the person is holding. It is probably fair to say that a 'white' South African citizen has the right to the same smooth passage as a European citizen, and perhaps even smoother than the passage of someone of Nigerian origins, for example, who is holding a European passport.

10. And with the current measures of securitisation, the deployment of some of these technologies (such as biometrics) is spreading into the entire 'circulating body'. Yet, there remains the decisive difference in terms of the meaning and the effect these technologies have on each category of this circulating body.

11. As with the 58 Chinese immigrants who died inside an airtight truck at the port of Dover in 2000 and the 19 immigrants who died inside a tractor-trailer in Houston, US in 2003. Here and in condemning the politics of borders, we should not forget to also condemn the kleptocratic authorities or ill-governed territories from which these unfortunate people are fleeing or being trafficked, nor should we forget to condemn the exploitation and manipulation inflicted by traffickers on these people. Yet one can maintain the argument that the increasingly stringent policies of borders in the West contribute significantly to creating more opportunities for exploitation. In fact, the intricate and cyclical relationship between these factors deserves a separate discussion in its own right, something that admittedly goes beyond the scope of the present study.

12. For example, 'In Brescia, [...] taxis with African drivers or passengers are stopped by police for verification of papers, to grant the right to move from one part of town to another' (Raj, 2005: 4).

13. In fact, the 'image' of being a border is so powerful that, in Mexico, it has been turned into a theme of 'entertainment', making 'misery and misfortune [...] a potentially profitable activity' (Rose, 1999: 260). Visitors of the Mexican theme park Parque EcoAlberto are now able to 'simulate' illegal border-crossing during the 'caminata nocturana' experience. This 'thrilling' five-hour adventure trek includes going through a drainage tunnel, under barbered wire and getting arrested by border guards. Of course, these 'paying' visitors are exempt from experiencing heat strokes, hypothermia, dehydration, drowning, and the like that cost the life of several hundred 'real' border-crossers each year (see http://www.parqueecoalberto.com.mx/caminata.html). More recently, the American computer firm Owlchemy has created the game Smuggle Truck for iPhone and iPad users in which players smuggle illegal immigrants over from Mexico. The game has been condemned as 'tasteless' and was eventually rejected by Apple in April 2011 (see http://www.dailymail.co.uk/news/article-1354221/Border-smuggling-game-stirs-controversy-iPhone-players-drive-immigrants-Mexcio.html).

14. Again, it is easy to simply direct the outrage at the human traffickers, as was the case with the deaths of the Chinese cockle pickers in Lancashire's Morecambe Bay. But one must recognise the wider and intertwined economic and political dimensions that lead to sustaining the cycle of exploitation and cheap labour (see http://news.bbc.co.uk/1/hi/england/lancashire/ 3464203.stm).

15. Which is argued to be 'draining the skilled and educated of the Third World' (Yuval-Davis et al., 2005: 520); the so-called phenomenon of '*l'exode des cerveaux*' (brain drain).

16. This, with the cooperation of 'the EU's "circle of friends," by conditioning economic aid to the permission to have police and immigration activities inside each of these countries' (Bigo, 2005a: 6). See, for example, the bilateral relations between Morocco and Spain concerning the issue of 'illegal immigration'. Morocco's cooperation with Spain is based on a string of incentives such as fishing agreements and support for Morocco's bid to obtain advanced status with the EU.

17. To use Walters' (2004: 255) apt metaphor with regard to immigration control.

18. Bio-sovereignty can also be defined as the bridging between biopower (Foucault) and sovereign power (Agamben).

19. Of course, one may raise the objection here that humanitarian interventions in the issue of asylum are not always state-run, nor is the state's decisionist monopoly always sustainable, since there are various other actors, such as NGOs, human rights groups and religious organisations, who partake of such humanitarian operations. However, and as many theorists have argued, there still remains the fact that these organisations share with the state the same 'object' of intervention, that is to say, the *life* of those seeking protection. In this respect, one should not be dismissive of Agamben's (1998: 133) argument that humanitarian organisations, despite themselves, 'maintain a secret solidarity with the very powers they ought to fight' (see also Hardt and Negri, 2000: 36–7).

20. Though not identical ones.

21. See also Didier Bigo (2005a).

22. After all, no one ever *punishes* policy-makers (and those who function as 'administrators without responsibility', to put it in Arendt's words) for implementing life-threatening policies. These liability-free (techno)bureaucratic policies are epitomes of what Arendt sees as the 'rule by Nobody that is clearly the most tyrannical of all since there is no one left who could even be asked to answer for what is being done […] no men, neither one nor the best, neither the few nor the many, can be held responsible' (Arendt, 1969: 233).

23. Claiming benefits under multiple identities.

24. At abandoned holiday camps and military bases (Telegraph, 29 October 2001).

25. Instead of the previous 'voucher system'.

26. Nevertheless, detention centres still remain in operation for holding new arrivals, failed asylum seekers and irregular migrants. Also, accommodation centres did not replace completely the controversial 'dispersal system' whereby asylum seekers are sporadically placed in different communities.

The system has been accused for leading to a series of disturbances includ-ing the murder of a Kurdish asylum seeker in Glasgow in 2001 (Telegraph, 29 October 2001).

27. In which exception is merely regarded as the *opposite* of the rule.
28. As in Agamben's reasoning.
29. In *Precarious Life*, Judith Butler (2004: 56) suggests a similar point vis-à-vis sovereignty in the following way: 'It is, of course, tempting to say that something called the "state," imagined as a powerful unity, makes use of the field of governmentality to reintroduce and reinstate its own forms of sovereignty. This description doubtless misdescribes the situation, however, since governmentality designates a field of political power in which tactics and aims have become diffuse, and in which political power fails to take on a unitary and causal form. But my point is that precisely because our histor-ical situation is marked by governmentality, and this implies, to a certain degree a loss of sovereignty, that loss is compensated through the resur-gence of sovereignty within the field of governmentality. Petty sovereigns abound, reigning in the midst of bureaucratic army institutions mobilized by aims and tactics of power they do not inaugurate or fully control. And yet such figures are delegated with the power to render unilateral decisions, accountable to no law and without any legitimate authority'.
30. This suggestion also claims that the Schmittian take on the notion of excep-tion (in terms of the sovereign suspension of the law) does not exhaust all other possible forms of its actualisation.
31. 'Exceptionalism' is not the same thing as 'exception' but refers to the 'rules of emergency and their tendency to become permanent' (Bigo, 2005a: 17).
32. Nancy also takes issue with the Schmittian sovereign decision. In *The Sense of the World* (1997: 93), he writes, 'Decision [...] does not take place for one alone or for two but for many, decides itself as a certain *in* of the in-common. Which one? Decision consists precisely in that *we* have to decide on it, in and for our world, and thus, first of all, to decide on the "we," on who "we" are, on how *we* can say "we" and can call ourselves *we*'. Hence, the importance of the question of community to which I shall return in a later chapter.
33. Not in the sense of the totality of a body (although new full-body scan machines are envisioned to do so, see, for instance, http://news.bbc.co.uk/ 1/hi/uk/8303983.stm), but that of all bodies equivalently.

3   Recombinant Identities: Biometrics and Narrative Bioethics

1. See also Agamben (1993).
2. See also Nancy (2000), especially his discussion on the notion of 'co-appearance' (we will return to Nancy's discussion in a later chapter).
3. For example, being assigned the identity of a refugee belongs to the sphere of the what, that is, an institutional identity *attribution* that (dis)qualifies the person as belonging to a certain category. What follows from this attribu-tion will have a bearing on the life experience of the person, on her 'story' and hence on her whoness, while *narrating* one's life as that of a refugee will

also inevitably affect the kind of attributions and status the person receives (especially in terms of rights, access, obligations, etc.); and one may also argue that the 'story of the refugee' would not come into being in the first place were it not for the existence of that bounded category of the citizen (which constitutes one of the contents of the 'what'). In such a context, the two formulations remain inextricably intertwined. They are both interwoven into the fabric of identity and 'happen' within a seemingly recursive movement, which contributes to the mutual transformation of the two and the forming of a continuum between 'what' and 'who'.

4. See Chapter 1.
5. See the discussion in the Introduction regarding Alterman's distinction between *biocentric* data (e.g. biometric data) and *indexical* data (e.g. social security number, driver's licence number).
6. I ought to point out here that, although there is much conceptual import from psychological-continuity theory and analytical philosophy in Ricoeur's thesis, Ricoeur does not entirely subscribe to their 'criteriological' distinctions, namely those of the bodily criterion and the psychological criterion (see Ricoeur, 1992: 128–9).
7. See also Ceyhan (2008: 116).
8. And here, we should stress again that the use of the body in the domain of identity/identification is unique neither to twenty-first century nor to biometric technology.
9. The production of the 'refugee identity' for example.
10. Even the new generation of biometric technology, which claims to be able to 'read the mind', remains unable to predict who someone is (see http://www.dailymail.co.uk/sciencetech/article-1060972/The-airport-security-scanner-read-mind.html).
11. Interestingly, for Jean-Luc Nancy, one way of *interrupting* such substantialist discourses (e.g. citizenship, individuality and community) is through 'literature' and 'writing', which bring to the fore the singularity of each and everyone, and resist forms of identitarianism and fusion (be they political, national, societal or otherwise). And, Ricoeur (1992: 115) describes literature as 'a vast laboratory [...] through which narrativity serves as a propaedeutic to ethics.'
12. Most of the current technological developments are geared towards this dimension of 'at distance'. Ironically, their performance is often measured and judged by how much distance they can flatten as well as how much distance they can guarantee and maintain. Some 'touch devices' are, in fact, designed to eliminate touch. Notice, for instance, next time you board a London bus and 'touch' your Oyster Travelcard, that there is no more need to address or even 'look' at the bus driver. Just 'scan and go', thus is the way!
13. See also 'Introduction' in Nelson (1997).
14. I am borrowing this concept from Megan Clinch.
15. Appropriating the title of Robert Musil's novel *The Man without Qualities*.
16. Although a painful one, given the circumstances that push one to seek asylum.
17. Here, suffering is not to be understood solely in a negative sense, but as an entire spectrum of experiences and affects including those of resistance, defiance and transgression (of borders and interiority, for instance).

18. I am intentionally using the phrase of 'the person seeking asylum' instead of 'asylum seeker', as the former denotes an 'action' whereas the latter is merely an identity ascription.

19. Jan Marta (1997: 206) also speaks of the notion of segregation when addressing some aspects of the physician–patient dynamics in relation to the issue of 'informed consent'.

20. See also Ricoeur's (2005) thorough discussion in *The Course of Recognition*.

21. This notion of 'singular plural' is discussed in more detail in a later chapter.

22. I borrow this phrase from Walter Benjamin (1968).

23. As is the case with 'myth', see discussion in Chapter 5.

24. I am referring here to 'the fact of exposing someone to death, increasing the risk of death for some people, or, quite simply, political death, expulsion, rejection, and so on' (Foucault, 2003 [1976]: 256), examples of which, as discussed in the previous chapter, can be found in the tragic deaths that are still taking place in the Strait of Gibraltar and the US–Mexican border.

25. I will take up this point in a later chapter while discussing Jean-Luc Nancy's philosophy.

4 Identity Securitisation and Biometric Citizenship

1. See discussion in the second section of Chapter 2.

2. Phishing is an online activity that uses fraudulent emails and spoof websites to gain unauthorised access to personal data including financial information (see www.antiphishing.org).

3. Although one has to be wary of the generalising aspect of such as statement as not 'all' market conditions are really 'borderless'.

4. Private not necessarily in the sense that these rights are granted by non-government entities but insofar as they delineate an exclusive space of flow that is separate from the rest and whose access depends on satisfying various neoliberal criteria. In the IRIS case, 'frequency' of travel is one of the (implicit) criteria guaranteeing legitimacy to enroll on the scheme. This correlation is itself implicated in the neoliberal rationality.

5. Although in reality, this is not always the case as some of these biometric entry style schemes tend to be 'national'. In the example of the IRIS system, for instance, the enrolled traveller has to queue up like everyone else when entering or exiting a 'foreign' border. It is only when coming back to the UK that he or she can use the IRIS programme to speed up the crossing.

6. Some may contend here that the current criteria for enrolling on the IRIS scheme do not require the person to be 'that' privileged and as such, this argument may rather be an overstatement. I would, however, maintain that the privileged side of this mobility scheme is most apparent when set against the background of the border-crossing conditions endured by asylum seekers and 'clandestine' migrants (see http://www.ukba.homeoffice.gov.uk/travellingtotheuk/Enteringtheuk/usingiris/registeriris/caniregisteriris/).

7. See also Adey's (2004) discussion regarding a similar scheme in the United States, namely the NEXUS system.

8. See also my previous discussion on 'recombinant identities'.

9. And also what is considered as the intentional use of freedom to 'abuse or attack freedom', particularly in the case of the 'war on terror'.

10. Epistemic in that it stems from the will to arrive at the 'absolute knowingness of the other' (Zylinska, 2005: 34) which, as Levinas (in ibid.) argues, 'amounts to grasping being out of nothing or reducing it to nothing, removing it from its alterity'.

11. For a more detailed discussion on the relationship between abjection and the politics of citizenship and immigration, see Tyler's (2010) 'Designed to fail: a biopolitics of British Citizenship' and Nyers' (2003) 'Abject Cosmopolitanism: the politics of protection in the anti-deportation movement'.

12. See 'Asylum protester sews up eyes', http://news.bbc.co.uk/l/hi/england/ 2939156.stm, and 'Sweden refugees mutilate fingers', http://news.bbc.co.uk/ 2/hi/europe/3593895.stm.

13. These acts are but some of the broad array of practices of resistance emerging alongside neoliberalism and in response to a growing politics of control. They demonstrate how mobility and citizenship are increasingly becoming a site of ongoing struggle and contestation. For a more detailed account on this, see, for instance, Squire (Ed.) (2011) *The Contested Politics of Mobility* and Papadopoulos et al. (2008) *Escape Routes: Control and Subversion in the 21st Century*.

14. In neoliberalism, just as freedom is reframed as that which pertains primarily to freedom of choice in terms of consumption and lifestyle, mobility and market opportunity, so too is the notion of responsibility. As Odysseos (2010: 753) argues, neoliberal freedom 'invokes individual responsibility in the sense of fending for oneself, that is, being self-sufficient'. In the case of refugees, asylum seekers and irregular migrants, freedom and responsibility form a peculiar continuum. Perceived as incapable of being self-sufficient and excluded from neoliberal freedoms, the sense of responsibility of these figures is also called into question by extension and in a way that reduces both freedom and responsibility to the mere question of whether one is included in and possesses the ability to execute neoliberal transactions.

15. In our case the community of 'legal travellers/residents'.

16. Waldby initially introduced the concept of biovalue in the context of human tissue and stem cell research. This term, however, is relevant to many other contexts whereby biology, capital and ethics are in interaction with each other.

17. See http://www.guardian.co.uk/world/2010/jan/01/racial-religious-groups-airport-checks and http://www.dailymail.co.uk/news/article-1240391/Body-scanners-approved-anger-ethnic-profiling.html.

18. Here I focus on race, but gender and class are also some of the important elements that warrant examination in relation to the politics of biometric identification.

19. Although in some other contexts, it is precisely this nexus that is at stake. See, for instance, the case of biometrics usage in Iraq and Afghanistan: http://www.usnews.com/articles/news/iraq/2008/05/ 01/the-us-army-ramps-up-biometrics-to-id-baghdad-residents.html; http:// www.wired.com/dangerroom/2007/08/fallujah-pics/; http://www.human

events.com/article.php?id=35735;   and   http://www.army.mil/-news/2009/
06/01/21940-biometrics-on-the-ground-and-in-the-dod/.
20.   See also the Northern Ireland Human Rights Commission's response to *The Path to Citizenship*.
21.   In 2005, the firm claimed that SDS-VR-1000 would undergo testing in some U.S airports. I could not find further proof of that. It is, of course, important to distinguish hype and marketing campaigns from what is implemented in reality. Nonetheless, these technological developments are interesting examples of the kind of techniques and rationalities that are currently emerging in the security field and its industry.
22.   So that it is not merely a matter of 'scan and go' but 'scan, detect, and then either let go or investigate further'.
23.   This 'virtual self' is not to be regarded as entirely virtual but as interwoven into the actuality of the flesh and blood individual, this, not only in terms of the abstract referential relationship between the virtual self and the actual individual but also in terms of the 'material' effects (including the emotional dimensions that are amplified during the occurrence of identity theft and fraud) of the former on the latter. Identity theft, as Whitson and Haggerty (2008: 575) argue 'represents the dramatic moment when the tensions, ironies an contradictions inherent in the relationship between the dividual and its human *doppelgänger* are most starkly revealed'.
24.   See details in a previous section.
25.   For instance, using the hand to protect the PIN number, when being typed, from being recoded by hidden cameras and using the bodily posture to protect these details from shoulder-surfing.
26.   See, for instance, the Identity Theft Resource Center's survey findings in 'Identity theft: the Aftermath' (2007), http://www.idtheftcenter.org/artman2/uploads/1/Aftermath_2007_20080529v2_1.pdf.
27.   See, for example, 'ITRC Fact Sheet 108 – Overcoming the Emotional Impact' (Identity Theft Resource Center, 2003), http://www.idtheftcenter.org/artman2/publish/v_fact_sheets/Fact_Sheet_108_Overcoming_The_Emotional_Impact.shtml.

5   Rethinking Community and the Political through Being-with

1.   'Until this day history has been thought on the basis of a lost community – one to be regained or reconstituted. The lost, or broken, community can be exemplified in all kinds of ways, by all kinds of paradigms: the natural family, the Athenian city, the Roman Republic, the first Christian community, corporations, communes, or brotherhoods – always it is a matter of a lost age in which community was woven of tight, harmonious, and infrangible bonds and in which above all it played back to itself, through its institutions, its rituals, and it symbols, the representation, indeed the living offering, of its own immanent unity, intimacy, and autonomy [...] at every moment in its history, the Occident has given itself over to the nostalgia for a more archaic community that has disappeared, and to deploring a loss of familiarity, fraternity and conviviality' (Nancy, 1991: 9–10).

2. 'Politically speaking the significance of the subject is that there is a relation to others only on the basis of an autonomous individual who preexists all communal formations and decides independently whether she or he enters sociality or not, whether he or she "signs" the social contract or not' (Devisch, 2000: 241).

3. So here, and as May (1997: 22) explains, totalitarianism/immanentism is used to refer to something 'wider than – although related to – the sense of the term when it is used to categorize a type of state governance. [It] refers to narrow constraints placed upon individual and social identity and behaviour rather than just a type of state [...] totalitarianism in that sense need not be a product of state totalitarianism [...] self-definition within narrowly defined parameters can itself be called [...] "totalitarianism." '

4. Although the issue of citizenship is present in other texts, notably in *The Sense of the World* (1997).

5. Although the *paradigm* of this modality, as Agamben contends, is not unique to the contemporary era.

6. We shall return to these notions of being-with and being-in-common later on.

7. From a speech by Michael Howard, the former Leader of the British Conservative Party.

8. It is as opposed to the singular stories we spoke about in Chapter 3. For the difference here is that myth does not share the features of these singular stories, that is, their inherent incompleteness, indefinitiveness, fragility and uncertain relation to their origins. Instead, myth is above all fixed, full, original speech that is, nevertheless, amenable to *interruption* by singularity itself.

9. Its simplicity derives from the kind of questions it engages with, namely and as we shall see, the questions concerning the meaning of our naked existence and 'the sheer banality of our contact (*cotoiment*) with the world and with others' (Critchley, 1999: 54).

10. Not necessarily in the traditional sense of the term. His is an empiricism of sense, of 'the polymorphy and polyphony of the banalities of common life' (Hutchens, 2005: 3), a phenomenology of 'the extremely humble layer of our everyday experience' (Nancy, in Critchley, 1999: 54).

11. See also the discussion in Chapter 3 relating to Cavarero's take on identity.

12. A simple analogy would be that of the case when one retreats after having an illness. That space of retreat can provide the person with an opportunity to rethink anew what the essence of health is and what being healthy means.

13. That is not to say that Nancy ignores the reality of power relations altogether. As the following indicates, 'I do not wish to neglect the sphere of power relations: we never stop being caught up in it, being implicated in its demands. On the contrary, I seek only to insist on the importance and gravity of the relations of force and [...] struggles of the world at the moment when a kind of broadly pervasive democratic consensus seems to make us forget that "democracy," more and more frequently, serves only to assure a play of economic and technical forces' (Nancy, 1991: 37).

14. For a more detailed comparison between Nancy's and Schmitt's approaches towards the political, see Levett's (2005) 'Taking Exception to Community (between Jean-Luc Nancy and Carl Schmitt)'.

15. See also Luszczynska's (2009) discussion.
16. One might ask here (or even object), is this not spirituality (with a Buddhist streak to be more precise) dressed up as the political? Perhaps it is so, and if that is the case, it is certainly a case of spirituality without a (holy) spirit and a *sense* of the political that is about being rather than organising, calculating and managing.
17. See also discussion in Chapter 3.

Bibliography

Aas, K.F. (2006) ' ' "The body does not lie" ': Identity, risk and trust in technocul-ture', *Crime Media Culture*, vol. 2, no. 2, pp: 148–153.

Adey, P. (2004) 'Secured and sorted motilities: Examples from the airport', *Surveillance & Society*, vol. 1, no. 4, pp: 500–519.

Adey, P. (2009) 'Facing airport security: affect, biopolitics, and the preemptive securitisation of the mobile body', *Environment and Planning*, vol. 27, no. 2, pp: 274–295.

Agamben, G. (1993) *The Coming Community*, Minneapolis: University of Minnesota Press.

Agamben, G. (1998) *Homo Sacer: Sovereign Power and Bare Life*, California: Stanford University Press.

Agamben, G. (2002) 'What is a Paradigm?', http://www.egs.edu/faculty/agamben/agamben-what-is-a-paradigm-2002.html.

Agamben, G. (2004) 'No to Bio-Political Tattooing', http://www.ratical.org/ratville/CAH/totalControl.html.

Agar, J. (2001) 'Modern Horrors: British Identity and Identity Cards', in Caplan, J. and Torpey, J. (eds), *Documenting Individual Identity*, New Jersey: Princeton University Press.

Alterman, A. (2003) ' "A piece of yourself": Ethical issues in biometric identifica-tion', *Ethics and Information Technology*, vol. 5, no. 3, pp: 139–150.

Amoore, L. (2006) 'Biometric Borders: Governing mobilities in the war on terror', *Political Geography*, vol. 25, no. 3, pp: 336–351.

Aradau, C. (2001) 'Migration: The Spiral of (In)Security', http://venus.ci.uw.edu.pl/~rubikon/forum/claudia1.htm.

Aradau, C. and van Munster, R. (2005) 'Governing Terrorism and the (non-) pol-itics of risk', University of Southern Denmark Political Science Publications, http://www.sam.sdu.dk/politics/publikationer/05Rens11.pdf.

Arendt, H. (1958) *The Human Condition*, Chicago: Chicago University Press.

Arendt, H. (1966) *The Origins of Totalitarianism*, London: Harcourt Brace Jovanovitch.

Arendt, H. (1969) *On Violence*, New York: Harcourt Brace.

Arras, J.D. (1997). 'Nice Story, But So What? Narrative and Justification in Ethics', in Nelson, H.L. (ed.), *Stories and Their Limits: Narrative Approaches to Bioethics*, New York: Routledge.

Athanasiou, A. (2003) 'Technologies of humanness, aporias of biopolitics, and the cut body of humanity', *Differences*, vol. 14, no. 1, pp: 125–162.

Atkins, K. (2000) 'Personal identity and the importance of one's own body: A response to Derek Parfit', *International Journal of Philosophical Studies*, vol. 8, no. 3, pp: 329–349.

Atkins, K. (2004) 'Narrative identity, practical identity and ethical subjectivity', *Continental Philosophy Review*, vol. 37, no. 3, pp: 341–366.

Aus, J.P. (2003) 'Supranational Governance in an "Area of Freedom, Security and Justice": Eurodac and the Politics of Biometric Control', http://www.sussex.ac.uk/sei/documents/wp72.pdf.

Balibar, E. (1991) 'Is There a "Neo-Racism"?', in Balibar, E. and Wallerstein, I. (eds), *Race, Nation, Class: Ambiguous Identities*, London: Verso.

Balibar, E. (1995) 'Culture and Identity (Working Notes)', in Rajchman, J. (ed.), *The Identity in Question*, New York: Routledge.

Balibar, E. (2002) *Politics and the Other Scene*, London: Verso.

Balibar, E. (2004) *We, the People of Europe? Reflections on Transnational Citizenship*, Oxford: Princeton University Press.

Baudrillard, J. (1983) *Simulations*, New York: Semiotext[e].

Baudrillard, J. (1990) *Seduction*, Basingstoke: Macmillan Education.

Bauman, Z. (2002) *Society Under Siege*, London: Polity.

Bauman, Z. (2004) *Identity*, Cambridge, UK: Polity Press.

Benjamin, W. (1968) *Illuminations*, New York: Schocken Books.

Bhabha, H. (1994) *The Location of Culture*, London: Routledge.

Bhandar, D. (2004) 'Renormalizing citizenship and life in fortress North America', *Citizenship Studies*, vol. 8, no. 3.

Bigo, D. (2005a) 'Globalized (In)Security: The Field and the Ban-Opticon', http://www.wmin.ac.uk/sshl/pdf/CSDBigo170106.pdf.

Bigo, D. (2005b) 'Exception and Ban: Discussing the "State of Exception"', http://ciph.org/fichiers_pdfdivers/Interventions_2.pdf.

Bigo, D. (2006a) 'Security, Exception, Ban and Surveillance', in Lyon, D. (ed.), *Theorising Surveillance: The Panopticon and Beyond*, Devon: Willan Publishing.

Bigo, D. (2006b) 'Globalized (In)Security: The Field and the Ban-Opticon', in Bigo, D. and Tsoukala, A. (eds), *Illiberal Practices of Liberal Regimes – The (In)Security Games*, Paris: L'Hartmattan.

Bigo, D. and Tsoukala, A. (eds). (2006) *Illiberal Practices of Liberal Regimes – The (In)Security Games*, Paris: L'Hartmattan.

Bigo, D., Walker, R.B.J., Carrera, S., and Guild, E. (2007) 'The Changing Landscape of European Liberty and Security', http://www.leeds.ac.uk/jmce/chalMTR.pdf.

Blair, T. (2006) 'PM Defends ID Cards Scheme for The Daily Telegraph', http://www.pm.gov.uk/output/Page10360.asp.

Blunkett, D. (2004) 'Identity Cards Speech', http://www.ippr.org.uk/uploadedFiles/events/Blunkett%20ID%20speech.pdf.

Bogard, M. (2006) 'Surveillance Assemblages and Lines of Flight', in Lyon, D. (ed.), *Theorising Surveillance: The Panopticon and Beyond*, Devon: Willan Publishing.

Bolter, J. and Grusin, R. (1999) *Remediation: Understanding New Media*, Cambridge: MIT Press.

Bos, R.T. (2005) 'Giorgio Agamben and the community without identity', *The Editorial Board of the Sociological Review*, pp: 16–29.

Braun, B. (2007) 'Biopolitics and the molecularization of life', *Cultural Geographies*, vol. 14, no. 6, pp: 6–28.

Brody, H. (1997) 'Who Gets to Tell the Story? Narrative in Postmodern Bioethics', in Nelson, H.L. (ed.), *Stories and Their Limits: Narrative Approaches to Bioethics*, New York: Routledge.

Brouwer, E.R. (2002) 'Eurodac: Its limitations and temptations', *European Journal of Migration and Law*, vol. 4, no. 2, pp: 231–247.

Bunyan, T. (2005) 'EU: Report on Biometrics Dodges the Real Issues', http://www
.statewatch.org/news/2005/mar/17eu-biometric-report.htm.

Burke, A. (2002) 'Aporias of Security', *Alternatives*, vol. 27, no. 1, pp: 1–27.

Butler, J. (1993) *Bodies That Matter: On the Discursive Limits of "Sex"*, New York: Routledge.

Butler, J. (2004) *Precarious Life*, London: Verso.

Butler, J. (2005) *Giving an Account of Oneself*, New York: Fordham University Press.

Byrne, L. (2007) 'Securing Our Identity: A 21st Century Public Good', http://
www.ips.gov.uk/cps/rde/xchg/SID-47DA2A28-4960BC98/ips_live/hs.xsl/225
.htm?advanced=&searchoperator=&searchmodifier=&verb=&search_date
_from=&search_date_to=&stage=&search_event_subject=&search_category
=&search_query=&search_scope=&search_group=&varChunk.

Caldwell, A. (2004) 'Bio-Sovereignty and the Emergence of Humanity', http://
muse.jhu.edu/journals/tae/v007/7.2caldwell.html.

Caplan, J. and Torpey, J. (eds). (2001) *Documenting Individual Identity*, New Jersey: Princeton University Press.

Carblanc, A. (2009) 'Human Rights, Identity and Anonymity: Digital Identity and Its Management in e-Society', in Mordini, E. et.al. (eds), *Identity, Security and Democracy*, Amsterdam: IOS Press.

Cavarero, A. (2000) *Relating Narratives: Storytelling and Selfhood*, London and New York: Routledge.

Caygill, H. (1997) 'The Shared World – Philosophy, Violence, Freedom', in Sheppard, D. (ed.), *On Jean-Luc Nancy: The Sense of Philosophy*, London: Routledge.

Ceyhan, A. (2008) 'Technologization of security: Management of uncertainty and risk in the age of biometrics', *Surveillance and Society*, vol. 5, no. 2, pp: 102–123.

CIFAS (2007) 'Identity Fraud', http://www.cifas.org.uk/default.asp?edit_id
=566-56.

CIFAS (2009) 'The Anonymous Attacker', http://www.cifas.org.uk/download/
The_Anonymous_Attacker_CIFAS_Special_Report.pdf.

Cohen, S. (2003) *No One Is Illegal: Asylum and Immigration Control, Past and Present*, Staffordshire: Trentham Books Ltd.

Cole, S.A. (2003) 'Fingerprint Identification and the Criminal Justice System: Historical Lessons for the DNA Debate', http://www.hks.harvard.edu/dnabook/
Simon_Cole_(4)_2-24-03.doc.

Collins, A. (2007) 'Introduction: What Is Security Studies?', in Cillins, A. (ed.), *Contemporary Security Studies*, New York: Oxford University Press.

Cooper, M., Goffey, A. and Munster, A. (2005) 'Biopolitics for Now', vol. 7, http://
www.culturemachine.net/.

Côté-Boucher, K. (2008) 'The diffuse border: Intelligence-sharing, control and confinement along Canada's smart border', *Surveillance and Society*, vol. 5, no. 2, pp: 142–165.

Coward, M. (1999) 'Unworking the Fictions of Citizenship: The Post-National Democratic Community', http://www.sussex.ac.uk/Users/mpc20/
pubs/unworking.doc.

Cowen, D. and Gilbert, E. (2008) *War, Citizenship, Territory*, New York: Routledge.

Cresswell, T. (2006). *On the Move: Mobility in the Modern Western World*, London: Routledge.

Cresswell, T. and Merriman, P. (eds). (2008) *Geographies of Mobility: Practices, Spaces, Subjects*, London: Ashgate.

Critchley, S. (1999) 'With Being-with? – Notes on Jean-Luc Nancy's Rewriting of *Being and Time*', http://www.secure.pdcnet.org/8525737F005882F3/file/DC77E14197DAE12F852573A30068637E/$FILE/studpracphil_1999_0001_0001_0059_0073.pdf.

Daily Mail (24 September 2008) 'The Airport Security Scanner That Can Read Your Mind', http://www.dailymail.co.uk/sciencetech/article-1060972/The-airport-security-scanner-read-mind.html.

Dallmayr, F. (1997) 'An Inoperative Community? Reflections on Nancy', in Sheppard, D. (eds), *On Jean-Luc Nancy: The Sense of Philosophy*, London: Routledge.

Dean, M. (2004) 'Four theses on the powers of life and death', *Contretemps*, vol. 5, pp: 16–29.

de Goede, M. and Randalls, S. (2009) 'Precaution, pre-emption: Arts and technologies of the actionable future', *Environment and Planning*, vol. 27, no. 5, pp: 859–878.

Deleuze, G. (1992) 'Postscript on the Societies of Control', http://pdflibrary.files.wordpress.com/2008/02/deleuzecontrol.pdf.

Delsol, R. (2008) 'Ethnic Profiling, ID Cards and European Experience', http://www.nihrc.org/dms/data/NIHRC/attachments/dd/files/104/Ethnic_Profiling_ID_and_european_experiences.pdf.

Derrida, J. (1976) *Of Grammatology*, Baltimore: Johns Hopkins University Press.

Devisch, I. (2000) 'A trembling voice in the desert: Jean-Luc Nancy's rethinking of the space of the political', *Journal for Cultural Research*, vol. 4, no. 2, pp: 239–255.

Diken, B. (2004) 'From refugee camps to gated communities: Biopolitics and the end of the city', *Citizenship Studies*, vol. 8, no. 1, pp: 83–106.

Diken, B. and Laustsen, C.B. (2005) *The Culture of Exception: Sociology Facing the Camp*, New York: Routledge.

Dillon, M. (2005) 'Cared to death: The biopoliticised time of your life', *Foucault Studies*, vol. 1, no. 2, pp: 37–46.

Diprose, R. (2002) *Corporeal Generosity: On Giving with Nietzsche, Merleau-Ponty, and Levinas*, New York: SUNY.

Diprose, R., Stephenson, N., Mills, C., Race, K., Hawkins, G. (2008) 'Governing the future: The paradigm of prudence in political technologies of risk management', *Security Dialogue*, vol. 39, no. 2–3, pp: 267–288.

Edkins, J. (2005) 'Exposed singularity', *Journal for Cultural Research*, vol. 9, no. 4, pp: 359–386.

Elden, S. (2002) 'Plague, Panopticon, Police', *Surveillance and Society*, vol. 1, no. 3, pp: 240–253.

Emmers, R. (2007) 'Securitization', in Cillins, A. (ed.), *Contemporary Security Studies*, New York: Oxford University Press.

Engin, F.I. (2004) 'The Neurotic Citizen', *Citizenship Studies*, vol. 8, no. 3, pp: 217–235.

European Commission (2005a) 'Biometrics at the Frontiers: Assessing the Impact on Society', http://europa.eu.int/comm/justice_home/doc_centre/freetravel/doc/biometrics_eur21585_en.pdf.

European Commission (2005b) 'EURODAC Guarantees Effective Management of the Common European Asylum System', http://www.libertysecurity.org/article429.html.

European Union (2006) 'Eurodac a European Union-wide Electronic System for the Identification of Asylum-seekers', http://ec.europa.eu/justice_home/fsj/asylum/identification/fsj_asylum_identification_en.htm.

Faulks, K. (2000) *Citizenship*, London: Routledge.

Fichte, J.G. (1889 [1796]) *The Science of Rights*, London: Trubner & Co.

Fimyar, O. (2008) 'Using Governmentality as a Conceptual Tool in Education Policy Research', *Educate Journal*, http://www.educatejournal.org/index.php?journal=educate&page=article&op=viewFile&path%5B%5D=143&path%5B%5D=157.

Finn, J. (2005) 'Photographing fingerprints: Data collection and state surveillance', *Surveillance and Society*, vol. 3, no. 1, pp: 21–44.

Fleck, L. (1979) *Genesis and Development of a Scientific Fact*, Chicago: University of Chicago Press.

Foucault, M. (1975) *Discipline and Punish*, London: Allen Lane.

Foucault, M. (1979) *History of Sexuality: The Will to Knowledge*, UK: Allen Lane.

Foucault, M. (2003 [1976]) 'Society Must Be Defended', in Bertani, M. and Fotana, A. (eds), *Society Must Be Defended: Lectures at the College de France 1975–1976*, New York: Plagrave Macmillan.

Foucault, M. (2008 [1979]) *The Birth of Biopolitics: Lectures at the College de France 1978–1979*, New York: Palgrave Macmillan.

Frank, A.W. (1997). 'Enacting Illness Stories: When, What, and Why', in Nelson, H.L. (ed.), *Stories and Their Limits: Narrative Approaches to Bioethics*, New York: Routledge.

Fuller, G. (2003a) 'Perfect match: Biometrics and body patterning in a networked world'. *Fibre Culture*, vol. 1, no. 1, http://one.fibreculturejournal.org/fcj002.

Fuller, G. (2003b) 'Life in transit: Between airport and camp', *Borderlands e-Journal*, vol. 2, no. 1, http://www.borderlands.net.au/vol2no1_2003/fuller_transit.html.

Gandy, O.H. (1993) *The Panoptic Sort: A Political Economy of Personal Information*, Boulder, Colo: Westview.

Gates, K. (2005) 'Biometrics and post-9/11 technostalgia', *Social Text*, vol. 23, no. 2, pp: 35–53.

Genet, J. (1965) *The Thief's Journal*, Middlesex: Penguin Books Ltd.

Giddens, A. (1991) *Modernity and Self Identity*, Cambridge, UK: Polity Press.

Gilbert, E. (2007) 'Leaky borders and solid citizens: Governing security, prosperity and quality of life in a North American partnership', *Antipode*, vol. 39, no. 1, pp: 77–98.

Greenspan, H. (2003) 'Listening to Holocaust Survivors: Interpreting a repeated story', in Josselson, R. (ed.), *Up Close and Personal: The Teaching and Learning of Narrative Research*, Washington: American Psychological Association.

Groebner, V. (2001) 'Describing the Person, Reading the Signs in Late Medieval and Renaissance Europe: Identity Papers, Vested Figures, and the Limits of Identification, 1400–1600', in Caplan, J. and Torpey, J. (eds), *Documenting Individual Identity*, New Jersey: Princeton University Press.

Hacking, I. (1992) ' "Style" for historians and philosophers', *Studies in the History and Philosophy of Science*, vol. 23, no. 1, pp: 1–20.

Haggerty, K.D. and Ericson, R.V. (2000) 'The surveillant assemblage', *British Journal of Sociology*, vol. 51, no. 4, pp: 605–622.

Haimes, E. (2002) 'What can the social sciences contribute to the study of ethics? Theoretical, empirical and substantive considerations', *Bioethics*, vol. 16, no. 2, pp: 277–298.

Hall, S. (1993) 'The Question of Cultural Identity', in Hall, S. McGrew, T., and Held, D. (eds), *Modernity and Its Futures*, Trowbridge: Polity Press.

Hardt, M. and Negri, A. (2000) *Empire*, Cambridge: Cambridge Press.

Hayter, T. (2000) *Open Borders: The Case Against Immigration Controls*, London: Pluto Press.

Hazelton, L. (2008) 'The Airport Security Scanner that Can Read Your Mind', *Daily Mail*, September 24, http://www.dailymail.co.uk/sciencetech/article-1060972/The-airport-securityscanner-read-mind.html.

Hazlewood, P. (2006) *Britain Is Becoming 'Big Brother' Society*, Agence France Presse. http://www.netcharles.com/orwell/articles/britain-becoming-surveillance-society.htm.

Hedgecoe, A.M. (2004) 'Critical bioethics: Beyond the social science critique of applied ethics', *Bioethics*, vol. 18, no. 2, pp: 120–143.

Heidegger, M. (1962) *Being and Time*, Oxford: Blackwell.

Heidegger, M. (1977) *The Question Concerning Technology, and Other Essays*, New York: Harper and Row.

Heintz, C. and Origgi, G. (2004) 'Rethinking Interdisciplinarity: Emergent Issues', http://www.interdisciplines.org/interdisciplinarity/papers/11.

Hier, S.P. (2003) 'Probing the surveillant assemblage: On the dialectics of surveillance practices as processes of social control', *Surveillance & Society*, vol. 1, no. 3, pp: 399–411.

Home Office (2002a) 'Secure Borders, Safe Haven: Integration with Diversity in Modern Britain', http://www.archive2.official-documents.co.uk/document/cm53/5387/cm5387.pdf.

Home Office (2002b) 'IRIS Scheme Definition Document', http://www.ukba.homeoffice.gov.uk/sitecontent/documents/managingourborders/eborders/irisdownloads/schemedefinitiondocument.pdf.

Home Office (2003) 'Identity Cards: The Next Steps', http://www.homeoffice.gov.uk/docs2/identity_cards_nextsteps_031111.pdf.

Home Office (2004) 'Identity Cards Bill', http://www.publications.parliament.uk/pa/cm200506/cmbills/009/2006009.pdf.

Home Office (2006a) 'Identity Cards Act 2006', http://www.opsi.gov.uk/acts/acts2006/en/ukpgaen_20060015_en_1.

Home Office (2006b) 'Fraud Act 2006', http://www.opsi.gov.uk/acts/acts2006/pdf/ukpga_20060035_en.pdf.

Home Office (2006c) 'Businessmen and Frequent Travellers to Benefit from New Immigration Technology', http://press.homeoffice.gov.uk/press-releases/new-immigration-technology?version=1.

Home Office (2006d) 'Immigration, Asylum and Nationality Act 2006', http://www.opsi.gov.uk/acts/acts2006/pdf/ukpga_20060013_en.pdf.

Home Office (2006e) 'Data Capture and Sharing Powers for the Border Agencies', http://www.homeoffice.gov.uk/error404?errorurl=http://imblive:8080/documents/ria-data-capture211005.

Home Office (2007) 'UK Borders Act 2007', http://www.opsi.gov.uk/acts/acts2007/pdf/ukpga_20070030_en.pdf.

Home Office (2008a) 'Everyone's Unique: Let us keep it that way', http://www.ips.gov.uk/cps/rde/xchg/ips_live/hs.xsl/index.htm.

Home Office (2008b) 'Identity Theft Leaflet', http://www.identitytheft.org.uk/downloads.asp.

Home Office (2008c) 'Security in a Global Hub: Establishing the UK's New Border Arrangements', http://www.cabinetoffice.gov.uk/media/cabinetoffice/corp/assets/publications/reports/border_review.pdf.

Home Office (2008d) 'Introducing Compulsory Identity Cards for Foreign Nationals', http://www.statewatch.org/news/2008/mar/uk-compulsory-id-for-foreign-nationals.pdf.

Home Office. (2008e) 'The Path to Citizenship: Next Steps to Reforming the Immigration System', *Next Steps*, http://www.bia.homeoffice.gov.uk/sitecontent/documents/aboutus/consultations/pathtocitizenship/pathtocitizenship?view=Binary.

Howard, M. (2005) 'Howard: Firm but Fair Immigration Controls', http://www.conservatives.com/tile.do?def=news.story.page&obj_id=119004&speeches=1.

Hurwitz, A. (1999) 'The 1990 Dublin convention: A comprehensive assessment', *International Journal of Refugee Law*, vol. 11, no. 4, pp: 646–677.

Hutchens, B.C. (2005) *Jean-Luc Nancy and the Future of Philosophy*, Bucks: Acumen.

Huysmans, J. (2006) *The Politics of Insecurity: Fear, Migration and Asylum in the EU*, London: Routledge.

Huysmans, J. and Buonfino, A. (2006) 'Politics of Exception & Unease: Immigration, Asylum and Insecurity in Parliamentary Debates on Terrorism in the UK', http://www.midas.bham.ac.uk/Politics%20of%20Exception%20%20Unease%20ISA%202006.pdf.

Identity Theft Resource Center (2003) 'ITRC Fact Sheet 108 – Overcoming the Emotional Impact', http://www.idtheftcenter.org/artman2/publish/v_fact_sheets/Fact_Sheet_108_Overcoming_The_Emotional_Impact.shtml.

Identity Theft Resource Center (2007) 'Identity Theft: The Aftermath', http://www.idtheftcenter.org/artman2/uploads/1/Aftermath_2007_20080529v2_1.pdf.

Independent (The) (2 November 2006) 'Big Brother Britain 2006: "We are Waking Up to a Surveillance Society All Around Us"', http://news.independent.co.uk/uk/crime/article1948209.ece.

Isin, E. (2002) 'Citizenship after Orientalism', http://www.yorku.ca/robarts/projects/lectures/pdf/rl_isin.pdf.

Isin, E. (2004) 'The neurotic citizen', *Citizenship Studies*, vol. 8, no. 3, pp: 217–235.

James, I. (2005) 'On interrupted myth', *Journal for Cultural Research*, vol. 9, no. 4, pp: 331–349

James, I. (2006) *The Fragmentary Demand: An Introduction to the Philosophy of Jean-Luc Nancy*, Stanford: Stanford University Press.

Jennings, W. (2007) 'At No Serious Risk? Border Control and Asylum Policy in Britain, 1994–2004', http://www.lse.ac.uk/collections/CARR/pdf/Disspaper39.pdf.

Kaluszynski, M. (2001) 'Republican Identity: Bertillonage as Government Technique', in Caplan, J. and Torpey, J. (eds), *Documenting Individual Identity*, New Jersey: Princeton University Press.

Kelly, C. (2010) 'The "Israelification" of Airports: High Security, Little Bother', http://www.thestar.com/news/world/article/744199—israelification-high -security-little-bother.

Kember, S. (2006) 'Doing Technoscience as ("New") Media', in Morley, D. and Curran, J. (eds), *Media and Cultural Theory*, London: Routledge.

Koslowski, R. (2003) 'Information Technology and Integrated Border Management', http://se2.dcaf.ch/serviceengine/Files/DCAF/29432/ichaptersection _singledocument/9EB60F9B-E5A2-4986-8CD5-F6122E493439/en/04_paper _Koslowski.pdf.

Kottman, P.A. (2000) 'Introduction', in Cavarero, A. (ed.), *Relating Narratives: Storytelling and Selfhood*, London: Routledge.

Kundnani, A. (2008) Experiences of "New Suspects Communities" in Britain', http://www.nihrc.org/dms/data/NIHRC/attachments/dd/files/104/ID _cards_Experiences_of_new_suspect_communities_by_Arun_Kundnani.pdf.

Lacoue-Labarthe, P. and Nancy, J-L. (1997) *Retreating The Political*, New York: Routledge.

Lemke, T. (2002) 'Foucault, Governmentality, and Critique', http://www .andosciasociology.net/resources/Foucault$2C+Governmentality$2C+and +Critique+IV-2.pdf.

Lentzos, F. (2006), 'Rationality, risk and response: A research agenda for biosecurity ', *BioSocieties*, vol. 1, no. 4, pp: 453–464

Leonardo, I.A. (2003) 'Machines made to measure: On the technologies of identity and the manufacture of difference', *Electronic Almanac*, vol. 11, no. 11.

Levett, N. (2005) 'Taking exception to community (between Jean-Luc Nancy and Carl Schmitt)', *Journal for Cultural Research*, vol. 9, no. 4, pp: 421–435.

Levi, M. and Wall, D. (2004) 'Technologies, security, and privacy in the post-9/11 European information society', *Journal of Law and Society*, vol. 31, no. 2, pp: 194–220.

Levinas, E. (1969) *Totality and Infinity*, Pittsburgh, PA: Duquesne University Press.

Lister, M., Dovey, J., Giddens, S., Grant, I. and Kelly, K. (eds). (2003) *New Media: A Critical Introduction*, London: Routledge.

LSE. (2005) 'The Identity Project: An Assessment of the UK Identity Cards Bill and its Implications', http://is.lse.ac.uk/idcard/identityreport.pdf.

Luszczynska, A. (2009) 'Nancy and Derrida: On ethics and the same (infinitely different) constitutive events of being', *Philosophy and Social Criticism*, vol. 35, no. 7, pp: 801–821.

Lyon, D. (2003) 'Technology vs "terrorism": Circuits of city surveillance since September 11th', *International Journal of Urban and Regional Research*, vol. 27, no. 3, pp: 666–678.

Lyon, D. (2004) 'Identity Cards: Social Sorting by Database', http://www.oii.ox .ac.uk/resources/publications/IB3all.pdf.

Lyon, D. (2007) 'National ID cards: Crime-control, citizenship and social sorting', *Policing*, vol. 1, no. 1, pp: 11–118.

Lyon, D. (2008) 'Biometrics, identification and surveillance', *Bioethics*, vol. 22, no. 9, pp: 499–508.

Lyon, D. (2009) *Identifying Citizens: ID Cards as Surveillance*, Cambridge: Polity Press.

Magnet, S. (2007) 'Are Biometrics Race-Neutral?', http://www.idtrail.org/index2 .php?option=com_content&do_pdf=1&id=689.

Maguire, M. (2009) 'The birth of biometric security', *Anthropology Today*, vol. 25, no. 2, pp: 9–14.

Marron, D. (2008) ' "Alter reality": Governing the risk of identity theft', *British Journal of Criminology*, vol. 48, no. 1, pp: 20–38.

Marta, J. (1997) 'Toward a Bioethics for the Twenty-First Century: A Ricoeurian Poststructuralist Narrative Hermeneutic Approach to Informed Consent', in Nelson, H.L. (ed.), *Stories and Their Limits: Narrative Approaches to Bioethics*, New York: Routledge.

May, T. (1997) *Reconsidering Difference: Nancy, Derrida, Levinas, and Deleuze*, Pennsylvania: Pennsylvania State University Press.

Migration and Law Network (2008) ' "The Path to Citizenship: Next Steps in Reforming the Immigration System": Response of the Migration and Law Network', http://www.kent.ac.uk/law/migration/documents/ Path%20to%20Citizenship.doc.

Miller, P. and Rose, N. (2008) *Governing the Present*, Cambridge: Polity Press.

Mills, C. (2004) 'Agamben's messianic politics: Biopolitics, abandonment and happy life', *Contretemps*, vol. 5, pp: 42–62.

Mills, C. (2010) 'Continental philosophy and bioethics', *Journal of Bioethical Inquiry*, vol. 7, no. 2, pp: 145–148.

Montello, M. (1997) 'Narrative Competence', in Nelson, H.L. (ed.), *Stories and Their Limits: Narrative Approaches to Bioethics*, New York: Routledge.

Mordini, E. and Ottolini, C. (2007) 'Body identification and medicine: Ethical and social considerations', *Ann Ist Super Sanita*, vol. 43, no. 1, pp: 51–60.

Mordini, E. and Petrini, C. (2007) 'Ethical and social implications of biometric identification technology', *Ann Ist Super Sanita*, vol. 43, no. 1, pp: 5–11.

Muller, B. (2004) '(Dis)Qualified bodies: Securitization, citizenship and "identity management" ', *Citizenship Studies*, vol. 8, no. 3, pp: 279–294.

Muller, B. (2010) *Security, Risk and the Biometric State: Governing Borders and Bodies*, London: Routledge.

Mythen, G. and Walklate, S. (2006) 'Criminology and Terrorism: Which Thesis? Risk Society or Governmentality?', *British Journal of Criminology*, vol. 46, no. 3, pp: 379–398.

Nancy, J-L. (1991) *The Inoperative Community*, Minneapolis: University of Minnesota Press.

Nancy, J-L. (1993) *The Experience of Freedom*, Stanford: Stanford University Press.

Nancy, J-L. (1997) *The Sense of the World*, Minneapolis: University of Minnesota Press.

Nancy, J-L. (2000) *Being Singular Plural*, Stanford: Stanford University Press.

Nancy, J-L. (2005) 'Interview: The Future of Philosophy', in Hutchens, B.C. (ed.), *Jean-Luc Nancy and the Future of Philosophy*, Bucks: Acumen.

Nelson, H.L. (1997) 'Introduction: How to Do Things with Stories', in Nelson, H.L. (ed.), *Stories and Their Limits: Narrative Approaches to Bioethics*, New York: Routledge.

Nelson, R.D. (1992) 'The sociology of styles of thought', *The British Journal of Sociology*, vol. 43, no. 1, pp: 25–54.

Norris, C. (2007) 'The intensification and bifurcation of surveillance in British Criminal Justice Policy', *European Journal of Crime Policy Research*, vol. 13, pp: 139–158.

Northern Ireland Human Rights Commission (2008a) 'Compulsory Identity Cards for Foreign Nationals: UK Borders Act 2007 Consultation Document', http://www.nihrc.org/dms/data/NIHRC/attachments/dd/files/100/Response _to_Compulsory_ID_Cards_for_Foreign_Nationals_(April_2008).doc.

Northern Ireland Human Rights Commission (2008b) 'The Path to Citizenship: Next Steps in Reforming the Immigration System, Response of the Northern Ireland Human Rights Commission', http://www.nihrc.org/dms/data/NIHRC/ attachments/dd/files/100/Response_to_Path_to_Citizenship_(May_2008).pdf.

Nyers, P. (2003) 'Abject cosmopolitanism: The politics of protection in the anti-deportation movement', *Third World Quarterly*, vol. 24, no. 6, pp: 1069–1093.

Nyers, P. (2004) 'Introduction: What's left of citizenship?', *Citizenship Studies*, vol. 8, no. 3, pp: 203–215.

Odysseos, L. (2010) 'Human rights, liberal ontogenesis and freedom: Producing a subject for neoliberalism?' *Millennium: Journal of International Studies*, vol. 38, no. 3, pp: 747–772.

Ong, A. (2006) 'Mutations in citizenship', *Theory, Culture & Society*, vol. 23, no. 2–3, pp: 499–505.

Papadopoulos, D., Stephenson, N. and Tsianos, V. (2008) *Escape Routes: Control and Subversion in the 21st Century*, London: Pluto Press.

Perera, S. (2002) 'What is a Camp ... ?', *Borderlands e-Journal*, vol. 1, no. 1, http:// www.borderlandsejournal.adelaide.edu.au/vol1no1_2002/perera_camp.html.

Petersen, A. (2003) 'Governmentality, critical scholarship, and the medical humanities', *Journal of Medical Humanities*, vol. 24, no. 3–4, pp: 187–201.

Pugliese, J. (2007) 'Biometrics, infrastructural whiteness, and the racialized zero degree of nonrepresentation', *Boundary 2*, vol. 34, no. 2, pp: 105–133.

Pugliese, J. (2010) *Biometrics: Bodies, Technologies, Biopolitics*, New York: Routledge.

Rabinow, P. (1996) *Essays on the Anthropology of Reason*, New Jersey: Princeton University Press.

Rabinow, P. (2003) *Anthropos Today: Reflections on Modern Equipment*, Princeton and Oxford: Princeton University Press.

Rabinow, P. and Rose, N. (2003) *Thoughts on the Concept of Biopower Today*, http://www.molsci.org/research/publications_pdf/Rose_Rabinow_Biopower _Today.pdf.

Raj, K.V. (2005) 'Refugee, Border-Camp: Implications of Recent Critiques of the State for Radical International Relations Theory', http://kiosk.polisci.umn.edu/ information/mirc/kiosk/oldsched/ScheduleSpring2005/MIRCRaj.pdf

Rajaram, P.R. and Grundy-Warr, C. (2004) 'The irregular migrant as homo sacer: Migration and detention in Australia, Malaysia, and Thailand', *International Migration*, vol. 42, no. 1, pp: 33–64.

Ricoeur, P. (1992) *Oneself as Another*, Chicago: The University of Chicago Press.

Ricoeur, P. (2005) *The Course of Recognition*, Massachusetts: Harvard University Press.

Rose, N. (1999) *Powers of Freedom: Reframing Political Thought*, Cambridge: Cambridge University Press.

Rose, N. (2000a) 'Government and Control', *British Journal of Criminology*, vol. 40, no. 2, pp: 321–339.

Rose, N. (2000b) 'Governing Liberty', in Ericson, R.V. and Stehr, N. (eds), *Governing Modern Societies*, Toronto: University of Toronto Press.

Rose, N. (2001) 'The politics of life itself', *Theory, Culture & Society*, vol. 18, no. 6, pp: 1–30.

Rose, N. (2006) *The Politics of Life Itself: Biomedicine, Power and Subjectivity in the Twenty-First Century*, Woodstock: Princeton University Press.

Rose, N. (2007) 'Molecular biopolitics, somatic ethics and the spirit of biocapital', *Social Theory & Health*, vol. 5, no. 1, pp: 3–29.

Rose, N. and Novas, C. (2000) 'Genetic risk and the birth of the somatic individual', *Economy and Society*, vol. 29, no. 4, pp: 485–513.

Rose, N. and Novas, C. (2002) 'Biological Citizenship', http://www.lse.ac.uk/collections/sociology/pdf/RoseandNovasBiologicalCitizenship2002.pdf.

Salter, M.B. (2008) 'When the exception becomes the rule: Borders, sovreignty, and citizenship', *Citizenhsip Studies*, vol. 12, no. 4, pp: 365–380.

Savirimuthu, A. and Savirimuthu, J. (2007) 'Identity theft and systems theory: The Fraud Act 2006 in perspective', *Scripted*, vol. 4, no. 4, pp: 436–461.

Schechtman, M. (1990) 'Personhood and personal identity', *The Journal of Philosophy*, vol. 87, no. 2, pp: 71–92.

Schmitt, C. (1950) *Ex Capitivitate Salus*, Cologne: Greven Verlag.

Schmitt, C. (1985) *Political Theology: Four Chapters on the Concept of Sovereignty*, Cambridge, MA: MIT Press.

Schmitt, C. (1996) *The Concept of the Political*, Chicago: University of Chicago Press.

Schwarzmantel, J. (2003) *Citizenship and Identity: Towards a New Republic*, London: Routledge.

Schwarzmantel, J. (2007) 'Community as communication: Jean-Luc Nancy and "being-in-common" ', *Political Studies*, vol. 55, no. 2, pp: 459–476.

Secomb, L. (2000) 'Fractured Community', *Hypatia*, vol. 15, no. 2, pp: 133–150.

Shoval, S. (2005) 'New Israeli System IDs Terrorists without Profiling', http://www.worldtribune.com/worldtribune/05/front2453545.904861111.html.

Smith, J. (2008) 'In Full: Smith ID Card speech', http://news.bbc.co.uk/1/hi/uk_politics/7281368.stm.

Sparke, M. (2006) 'A neoliberal nexus: economy, security and the biopolitics of citizenship on the border', *Political Geography*, vol. 25, no. 2, pp: 151–180.

Squire, V. (ed). (2011) *The Contested Politics of Mobility*, New York: Routledge.

Stalder, F. and Lyon, D. (2003) 'Electronic Identity Cards and Social Classification', in Lyon, D. (ed.), *Surveillance as Social Sorting: Privacy, Risk and Digital Discrimination*, Oxon: Routledge.

Stasiulis, D. (2004) 'Hybrid Citizenship and What's Left', *Citizenship Studies*, vol. 8, no. 3, pp: 295–303.

Stasiulis, D. (2008) 'The Migration-Citizenship Nexus', in Isin, E. (ed.), *Recasting the Social in Citizenship*, Toronto: University of Toronto Press.

Statewatch (2005) *Biometry and Electronic ID Card: Big Brother is Watching You*, http://www.statewatch.org/news/2005/nov/aedh-biometric-ID-card.pdf.

Stepnisky, J. (2007) 'The biomedical self: Hermeneutic considerations', *Social Theory & Health*, vol. 5, no. 3, pp: 187–207.

Strawson, G. (2004) 'Against Narrativity', *Ratio*, vol. 17, no. 4, pp: 428–452.

Telegraph (29 October 2001) 'Refugees to be housed in new centres', http://www.telegraph.co.uk/news/main.jhtml?xml=/news/2001/10/29/nasy29.xml.

Telegraph (30 October 2002) 'Identity Cards and Cash for All Asylum Seekers', http://www.telegraph.co.uk/news/main.jhtml?xml=/news/2001/10/30/nasy30.xml.

Thacker, E. (2004) *Biomedia*, Minneapolis, MN: University of Minnesota.

Thacker, E. (2005) 'Nomos, Nosos and Bios in the Body Politic', *Culture Machine*, vol. 7, http://culturemachine.tees.ac.uk/frm_f1.htm.

Todd, M. (1997) *Reconsidering Difference: Nancy, Derrida, Levinas, and Deleuze*, University Park, PA: Pennsylvania State University.

Torpey, J. (2000) *The Invention of the Passport: Surveillance, Citizenship and the State*, Cambridge: Cambridge University Press.

Tyler, I. (2010) 'Designed to fail: A biopolitics of British citizenship', *Citizenship Studies*, vol. 14, no. 1, pp: 61–74.

Valo, M. (2006) 'Biométrie – extrême fichage: Danger!', *Le Monde*, September 2, pp: 20–26.

van Asselt, M. and Vos, E. (2006) 'Precautionary principle and the uncertainty paradox', *Journal of Risk Research*, vol. 9, no. 4, pp: 313–336.

van der Ploeg, I. (1999a) 'The illegal body: '"Eurodac"' and the politics of biometric identification', *Ethics and Information Technology*, vol. 1, no. 4, pp: 295–302.

van der Ploeg, I. (1999b) 'Written on the body: biometrics and identity', *Computers and Society*, vol. 37, no. 1, pp: 37–44.

van der Ploeg, I. (2003a) 'Biometrics and the body as information: Normative issues of the socio-technical coding of the body', in Lyon, D. (ed.), *Surveillance as Social Sorting: Privacy, Risk and Digital Discrimination*, Oxon: Routledge.

van der Ploeg, I. (2003b) 'Biometrics and privacy: A note on the Politics of theorising technology', *Information, Communication & Society*, vol. 6, no. 1, pp: 85–104.

van der Ploeg, I. (2005a) *The Machine-Readable Body: Essays on Biometrics and the Informatization of the Body*, Maastricht: Shaker Publishing.

van der Ploeg, I. (2005b) 'The Politics of Biometric Identification: Normative Aspects of Automated Social Categorization', http://www.biteproject.org/documents/politics_of_biometric_identity%20.pdf.

van der Ploeg, I. (2009) 'Machine-Readable Bodies: Biometrics, Informatization and Surveillance', in Mordini, E. et.al. (eds), *Identity, Security and Democracy*, Amsterdam: IOS Press.

van Munster, R. (2005a) 'Logics of Security: The Copenhagen School, Risk Management and the War on Terror', University of Southern Denmark Political Science Publications, http://www.sam.sdu.dk/politics/publikationer/RensSkrift10.pdf.

van Munster, R. (2005b) 'The EU and the Management of Immigration Risk in the Area of Freedom, Security and Justice', University of Southern Denmark Political Science Publications, http://www.sdu.dk/~/media/Files/Om_SDU/Institutter/Statskundskab/Skriftserie/05Rens12%20pdf.ashx.

Waldby, C. (2002) 'Stem cells, tissue cultures and the production of biovalue'. *Health*, vol. 6, no. 3, pp: 305–323.

Walters, W. (2004) 'Secure borders, safe haven, demopolitics', *Citizenship Studies*, vol. 8, no. 3, pp: 237–260.

Walters, W. (2005). *Figuring Security: Notes on Power, In/Security and Territory*, http://www.qub.ac.uk/polproj/reneg/global_norms-papers/Walters_Figuring_Security.pdf.

Webster, F. (2000) 'Information, capitalism and uncertainty', *Information, Communication & Society*, vol. 3, no. 1, pp: 69–90.

Welch, R.V. and Panelli, R. (2007) 'Questioning community as a collective antidote to fear: Jean-Luc Nancy's "singularity" and "being singular plural"', *Area*, vol. 39, no. 3, pp: 349–356.

Whitson, R.J. and Haggerty, D.K. (2008) 'Identity theft and the care of the virtual self', *Economy and Society*, vol. 37, no. 4, pp: 572–594.

Williams, M. (2003) 'Words, images, enemies: Securitization and international politics', *International Studies Quarterly*, vol. 47, no. 4, pp: 511–531.

Wills, D. (2006) 'The United Kingdom Identity Card Scheme: Shifting Motivations, Static Technologies', in Bennett, J.C and Lyon, D. (eds), *Playing the Identity Card: Surveillance, Security and Identification in Global Perspective*, New York: Routledge.

Wood, M.D. and Firmino, R. (2010) 'Empowerment or repression? Opening up identification and surveillance in Brazil though a case of "identification fraud"', *Identity in the Information Society (IDIS)*, vol. 2, no. 3, pp: 297–317

Woodward, J.D. (1998) 'Biometrics: Identifying Law & Policy Concerns', http://www.cse.msu.edu/~cse891/Sect601/textbook/19.pdf.

Yuval-Davis, N Anthias, F. and Kofman, E. (2005) 'Secure borders and safe haven and the gendered politics of belonging: beyond social cohesion', *Ethnic and Racial Studies*, vol. 28, no. 3, pp: 513–535.

Žižek, S. (2002) *Welcome to the Desert of the Real*, London: Verso.

Zylinska, J. (2004) 'The universal acts, Judith Butler and the biopolitics of immigration', *Cultural Studies*, vol. 18, no. 4, pp: 524–539.

Zylinska, J. (2005) *The Ethics of Cultural Studies*, New York: Continuum.

Index

Note: Letter 'n' followed by locators refer to notes.

CPSIA information can be obtained at www.ICGtesting.com
Printed in the USA
BVOW04*0006311213

340536BV00003B/8/P